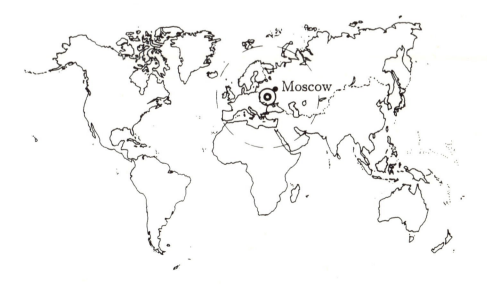

V. M. Chernousenko

Chernobyl

Insight from the Inside

With 124 figures and 47 table

Springer-Verlag
Berlin Heidelberg New York
London Paris Tokyo
HongKong Barcelona
Budapest

Dr. Vladimir M. Chernousenko
Institute for Theoretical Physics
Academy of Sciences of the Ukrainian SSR
Metrologicheskaya 14 b
SU-252130 Kiev
USSR

English by:

John G. Hine
124, St. Denis Road
Selly Oak
Birmingham B29 4LU
England

Photographs by:
Aleksandr I. Salmychin
ul. Marshala Zhukova 29-9-122
SU-252165 Kiev
USSR

Chapters 4, 5, 16,
and the Appendix by:

Dr. Natasha Aristov
Tiergartenstrasse 17
D-6900 Heidelberg 1
Federal Republic of Germany

ISBN 3-540-53698-1 Springer-Verlag Berlin Heidelberg New York
ISBN 0-387-53698-1 Springer-Verlag New York Berlin Heidelberg

Umschlagentwurf: Susanne Eisenschink, 6805 Hettesheim
Printed and bookbinding: F. W. Wesel, 7570 Baden-Baden
57/3140/543210 – Printed on acid-free paper

Dedication

To those who went through the Special Zone and to the people living in the radiation poisoned territory since 1986.

Preface

The Myths of Chernobyl,
and why I Wrote This Book

> Accidents at nuclear power stations happen.
> Between 1971 and 1986, in 14 countries,
> there were 152 accidents.
> *From information provided by the*
> *International Atomic Energy Agency*

> "It is not enough for a handful of experts
> to attempt the solution of a problem,
> to solve it and then to apply it.
>
> The restriction of knowledge
> to an elite group
> destroys the spirit of society
> and leads to its intellectual impoverishment."
> *Albert Einstein*

I must help to dispel some dangerous myths. After the Chernobyl disaster the polarization of public opinion with regard to nuclear power as a source of energy became even more pronounced than before. At one extreme are the representatives of the nuclear industry, who believe, despite the growing scale of nuclear reactor accidents that the development of nuclear power must continue at an unrelenting pace. At the other extreme, we find the "Greens" and others demanding the immediate shut-down of all operating reactors and a ban on future construction of nuclear power stations.

In my view, both extremes are too simplistic.

In the history of civilization there have been occasions when people could not refrain from developing hazardous industrial processes. Nuclear power is, of course, the most prominent example. We must accept the facts and our enormous energy requirements, but it is also important not to forget that attempts to economize on safety provisions in such hazardous industries result in increased risk. And these increased risks may result in terrible tragedy, even in disasters whose consequences exceed national boundaries. It is certainly true that a nuclear power station working safely without any accident, is ecologically one of the cleanest of all industrial plants. However, a single accident, like the one at Chernobyl, can negate all advantages for centuries to come.

If we feel, therefore, that we do not now have the ideas and resources to create absolutely effective radiation safeguards, then it would be better to call a halt today. Tomorrow may be too late.

Unfortunately, international public opinion has already been confused by the myths concerning the causes and scale of the Chernobyl disaster and its consequences for millions of people. Probably, the birth of these myths may be traced back to the articles published in the Soviet press in May 1986. The public was assured that "the heroes of Chernobyl" were "entering the Zone", "studying the situation", "bringing the reactor in Block 4 under control", "bringing the situation under control".

In reality, no means were available to bring the reactor or even the whole situation under control. The reactor was dead. Its radioactive core had already been torn apart by the explosion. Almost all the radioactivity it could release, had already been set free by May 10, 1986. Millions of Curies of radionuclides from the gutted reactor had been scattered across the face of the earth. A transnational nuclear disaster had already happened.

Then was not the time to save the nuclear power station, but to save the people – those living far beyond the boundaries of the 30-km Zone. However, the Government Commission charged with the rectification work (the "Liquidation of the Consequences of the Accident at the Chernobyl Nuclear Power Station" (LPA) "Likvidatsiya posledstvii avarii", stubbornly concentrated all its attention on the tiny 10-km Special Zone. It was into this small area that all the material resources were thrown, along with thousands of untrained and unprotected soldiers and reservists. The politicians had decided that the remaining three blocks of the station had to be brought back on-line, whatever the cost.

Hence, the birth of the myths of Chernobyl. A wave of disinformation swamped the Soviet press and then washed over the Western press.

A new wave of myths rippled out from the official report given by the Soviet (nuclear industry) representatives at the IAEA (International Atomic Engergy Agency) conference in August 1986. The report was full of vague formulations, unchecked data, and false conclusions as to the causes and scale of the disaster. Apparently, the IAEA was quite satisfied with it. All indications were that these conclusions were perfectly acceptable to top officials of the IAEA.

And the myths continue to be born. By the fifth anniversary of the tragedy, hundreds of books and articles had been written about Chernobyl. Of these, a number were written by people with no expert knowledge and are primarily emotional in character. Each of these seeks to throw light on one or another aspect of the disaster, usually concentrating on the first days or weeks.

The second category consists of works at least written by experts, but by experts who had no opportunity to personally check the situation in Chernobyl – or, perhaps, who had dropped by briefly, but only after 1987, and often only for the purpose of having their photograph taken with the Sarcophagus in the background. This picture would then be included in their book to emphasize the author's personal involvement in the events described. Most of their factual information about what happened, most of their "reliable" data are gleaned from the official Soviet press. Unfortunately, even in the period of perestroika and glasnost' the Soviet press is far from being objective, and the information it offers is far from being totally reliable. The truth about the tragedy has been concealed by very severe censorship. The world has been shown "Potemkin villages" full of happily resettled inhabitants. From the days of the Tsars to those of Stalin, Russia always knew how to erect a facade – and it is a well-known fact that the absence of reliable information, or of any information at all, creates ideal conditions for the birth of myths.

Myth 1 (for further details see Chapters 1 – 3): The design of the RBMK-1000 reactor is impeccable. It was the operating staff that caused the explosion.

Myth 2 (see Chapter 1): The radionuclides emitted from the shattered reactor represent only 3% of the full 192-ton charge of uranium.

Myth 3 (see Chapter 3): The partial technical changes to which the 15 RMBK-1000 reactors still in operation were subjected after the disaster have eliminated any danger of a second accident.

Myth 4 (see Chapter 1): Only 31 people died as a result of the accident and the cleanup operations.

Myth 5 (see Chapters 1 and 4): The concentration of attention on the 30-km Zone and on the decontamination and restarting of the three other reactors at Chernobyl was both permissible, given the radiation situation, and essential, given the need for electricity.

Myth 6 (see Chapter 4): Before the accident there already existed a proper, scientifically developed contingency plan backed up with all the necessary technical resources, and all designed to cope with a nuclear disaster of such a scale.

Myth 7 (see Chapter 5): With the completion of the Sarcophagus, rectification was essentially completed. The Sarcophagus is a tomb full of highly radioactive waste and is designed to last for 30 years. It is perfectly safe and presents no threat to people or environment.

Myth 8 (see Chapter 6): Work in ultra-high radiation fields was carried out with the help of robots. The men who entered such fields were equipped with the appropriate protective gear.

Myth 9 (see Chapter 7): When "rectifiers", after their service in the Zone, began to fall ill and die, the link between their illness and the time they had spent in radiation fields was recognized. They were given the necessary medicines and treatment while welfare assistance was also available to their families.

Myth 10 (see Chapter 8): People living outside the 30-km Zone, but in areas affected by the emission of radionuclides, were warned in good time of the danger which threatened them and were given iodine treatment. The civil defense system worked effectively.

Myth 11 (see Chapter 9): The "35-rem safe liftime dose" guideline for people living in contaminated areas (as proposed by Academician Il'yin and apparently supported by the leading officials of the IAEA) is scientifically sound and will not result in damage to health.

Myth 12 (see Chapters 1, 8, and 9): There is no reason to think that tens of thousands of children in the Ukraine, Byelorussia, and the Russian republic received radiation doses to their thyroid glands hundreds of times greater than the internationally permitted maximum dose.

Myth 13 (see Chapter 11): There is no reason to think that in the affected areas there is a continuing rise in the number of ailments caused by radiation, received either externally or internally through inhalation or contaminated food.

Myth 14 (see Chapters 1 and 11): The doses which people have received while living in contaminated areas will not have a genetic effect.

Myth 15 (see Chapter 12): The disaster will have no long-term impact on the environment, will not damage flora and fauna.

Myth 16 (see Chapter 13): The protective measures taken in the summer of 1986 in the 30-km Zone have prevented any leakage of radionuclides into the surface and ground waters.

Myth 17 (see Chapters 13 and 15): The radioactive pollution of the flood-lands of the river Pripyat and of the silt of the Kiev Reservoir does not create any threat to the Dniepr basin or, ultimately, to the Black Sea.

Myth 18 (see Chapter 14): The existing national and international norms created to protect the civilian population from radiation are scientifically sound and will prevent any damage to the health of future generations.

Myth 19 (see Chapter 15): In the event of another nuclear accident any-where in the world, scientific expertise and technical resources exist which are adequate for the task of decontaminating huge areas.

Myth 20 (see Chapters 14 and 15): The storage facilities for fuel waste and liquid waste at Chernobyl (and also at other Soviet nuclear power stations), the total content of which, in terms of radioactivity, amounts to more than 20 billion Curies, pose no threat to the world.

Myth 21 (see Chapters 1 and 15): The 800 – 1000 "tombs" which were dug in the (30-km) Zone to dump more than 500 million cubic meters of high-level and low-level radiactive debris present no danger to the world's aquifers.

From the preceding (incomplete) catalogue of myths it is seen that this book attempts to bring into the open true facts about the way the Chernobyl disaster has been handled – from April 1986 up to August 1991.

I deliberately chose to present the interviews in their original tone which is sometimes emotional. Along with personal messages conveying horror, pain, grief, disappointment, frustration, and anger, the reader will find reports and data presented in the language of science. These different messages provide an adequate overall picture of Chernobyl. They reflect the many facets of the tragedy.

Traditionally, science adopts the well-justified point of view that data can only be considered to be reliable if they can be and have been verified by several independent groups. The stress is on both "several" and "independent". This is not yet the case with much of the information presented here. The international scientific community has not yet had the opportunity to review all the measurements made by Soviet scientists. Consequently, only in a few cases was it possible to check the Soviet results against their own measurements. This book makes an important step towards establishing the truth as it contains a wealth of material that has not been presented previously and which is discussed without prejudice. It provides a basis for establishing an unbiased recognition of the facts.

The data presented concern all aspects of the accident, from contamination of soil and aquifers, technological and economical aspects to biological, medical, and psychological findings. Some chapters and sections contain material of a more technical nature and will appeal most to the specialist in the given area. The Appendix contains explanations of technical terms, some details and data providing the basis for some of the statements made in the main text, and thus also a guide for independent scientific studies.

The scientific data presented here are invaluable in assessing the real situation – in particular, in view of the policy of certain branches of the bureaucracy to hinder the publication of related data, in contradiction to President Gorbachev's policy of "glasnost". This policy is especially deplorable considering that the well-being of millions of people are concerned; it has already spauned poor communication between the population in the affected territories and the authorities. This dreadful situation has been clearly recognized by high-ranking officials of the Soviet Union who have already initiated countermeasures. This book provides additional support for such actions.

After five years of participation in the so-called rectification work I understand things ever more clearly. I know that it is not justifiable to speak about the "Rectification of the Consequences of the Accident". If one takes into account the scale and the degree to which an enormous number of peaceful people and a huge territory (in effect, our whole planet) have been affected, then one sees that the legacy of this catastrophe will continue to affect all of us for the rest of our lives.

Our primary goal should be to provide relief to the people who suffer from the catastrophe's direct consequences, and who are still living in the polluted territories. So far, only timid first steps in that direction have been taken; there are still very difficult times ahead of us. The noble, humanitarian participation of the international community is required for the good of all people.

Chernobyl – Paris – Heidelberg – London,   *Vladimir M. Chernousenko*
August 1991

Acknowledgements

This book could not have been written without the help of hundreds of people whom I have had the good fortune to meet during the period I have spent as an active participant in the rectification operation. Justice demands that their co-authorship be recognized. It is impossible to mention all of them, but I would like to mention explicitly at least some of those who supported me in this undertaking or suffered from my dedication to it. Among the latter are, in first place, my wife and my daughters; in addition to my already dreaded passion for physics, overnight they found me deeply involved in the problems of Chernobyl, which reduced family life almost to zero.

I want to express special gratitude to all the people with whom I worked in the Special Zone in that hot summer of 1986 and, in particular, to my friends Ye. Akimov, I. Akimov, V. Golubev, V. Golushchak, A. Gureyev, V. Dedov, Yu. Andreyev, G. Dmitrov, D. Vasilchenko, A. Nistryan, V. Omelchenko, Yu. Samoilenko, G. Nadyarnykh, V. Starodumov, A. Shimin, V. Chuchrin, V. Pshenichnykh, the colonels A. Nosach, A. Sontnikov, A. Saushkin, A. Grebenyuk, to generals K. Polukhin, N. Tarakanov, to the head of the special dosimetry unit A. Yurchenko, to V. Kulekin and A. Kuznetsov.

I also want to thank all the people who were on duty at Chernobyl on that tragic night of April 26, 1986. Many of them paid with their lives for their successful fight to prevent the explosion of the other three reactors of the Chernobyl nuclear power station. With some of them, I spent long evenings, when we were patients in Clinic No.6 in Moscow, reconstructing a second-by-second account of that terrible night. I would like to mention in particular O. Genrikh, V. Smagin, A. Nekhayev, A. Uskov, A. Tormozin, and A. Yurchenko.

While I was working on the book, I received an enormous amount of help from members of the "Chernobyl Union", in particular, from G. Lepin, N. Karpan, M. Melnikov, V. Tarasenko, V. Lomakin, V. Khalimchuk, and K. Sabadyr. Material that they prepared formed the basis by which I wrote Chapters 3, 9, 14, and 15. Material from the report written by A. Yadrikhinskii on the causes of the disaster was used partly in Chapter 1. Material from A. Nikitin's report has been used in Appendix C.

A word of acknowledgement is due to those people who have lived in the contaminated territories since 1986 and who have helped me to understand the situation there. I would like to mention T. Byelookaya, T. Grudnitskaya, V.

Yavlenko, A. Volkov, I. Makarenko, A. Mozhar, M. Sizonenko, S. Volynets, A. Nevmerzhitskii, You. Afonin, N. Nikitentko, A. Budko, and G. Mishchi.

For stimulating discussions of a number of technical questions relating to the rectification operation, I am indebted to Academicians A. Akhiezer, V. Baryakhtar, A. Davydov, R. Sagdeev, A. Sitenko, V. Trefilov, V. Kukhar, D. Grodzinskii, A. Dykhnya, V. Legasov, E. Sobotovich, to Doctors V. Novikov, Yu. Tsoglin, V. Shakhovtsov, G. Lisichenko, I. Sadolko, Yu. Okhovik, A. Selvestrov, M. Zheleznyak, L. Bolshov, Yu. Zaitsev, and V. Lisovenko, also to my fellow-scientists working at the "Collège de France" and at the research centers Karlsruhe, Jülich, Munich (where I am particularly grateful to Professor A.M. Kellerer) in Germany, and also at the University of Ravensburg, to the scientists of the Pugwash Movement, especially to Professors J. Rotblat and F. Calogero.

For unfailing support and for creating the conditions in which work on the book could be completed, I would like to express my gratitude to Springer-Verlag. For her support during my stay in Paris, I am very much indebted to Ms. M. Tovar. For their help during my stay in Heidelberg and for discussions at preliminary stages, and then for their active participation toward preparing the book for publication, I would like to expressly thank N. Aristov, E. Hefter, and S. von Kalckreuth from the Physics Editorial, who were later joined by Mr. J. Willis.

The last, but not the least of these acknowledgements goes to John Hine, whose excellent English and strong devotion enabled him to cope not just with a handwritten Russian original manuscript, but also to cope with a very tight deadline and to faithfully translate, nevertheless, my concerns and thoughts. May they be of benefit to the reader.

Foreword

From the Publisher:

Vladimir Mikhailovich Chernousenko was born May 12, 1941, in New York, a small village in Donetsk province. He studied physics at Kharkov State University, specializing in theoretical atomic physics, and graduated in 1965. He started his scientific career at the Ukrainian Academy of Sciences Institute of Physics in Kiev. Since 1971, he has worked at the Institute for Theoretical Physics of the Ukrainian Academy of Sciences in Kiev, where he earned his Ph.D. in theoretical physics in 1973. Since then, up to 1991, he has been the head of the Laboratory for Nonlinear Physics and Ecology. His scientific accumen is exceptionally diverse, as can be seen from his numerous publications (120 scientific papers and four monographs).

After the disaster at the Chernobyl nuclear power station in April 1986, he was invited by the Ukrainian Academy of Sciences to act as scientific director of their task force in Chernobyl. There he was concerned with conceiving the

appropriate rectification measures, such as the building of a containment around the shattered Block 4, the "Sarcophagus", devising special methods for work in unprecedentedly high radiation fields (reaching several thousand Roentgens), for decontamination and radiation protection measures, etc..

From May 1986 to January 1987, he worked in the Special Zone (10-km radius around the reactor) as Scientific Director of the Academy's Task Force for the Rectification of the Consequences of the Accident. He was a member of the Government Commission and took part in all its meetings. He is one of three authors of the secret report on the accident and the rectification measures prepared for the Soviet Government. Up to the beginning of 1991, he was Deputy Chairman of the Ukrainian Academy of Sciences Commission responsible for the Rectification of the Consequences of the Chernobyl Accident, and Scientific Director of the 30-km Zone.

Those who worked in the Special Zone, particularly around the exploded reactor were exposed to exceptionally high radiation doses. Therefore, the physicists, technicians, specialists for reactor technology, members of the army, and others working on the Task Force were rotated in 15-day shifts at the longest, or when they had received a dose of 25 rem. Only three persons stayed the entire time until the Sarcophagus was completed in November: Yu. M. Samoilenko, General Director of the Task Force, V. V. Golubev, Engineering Director, and the author of this book, V. M. Chernousenko, Scientific Director. It was only by withholding information from the medical authorities about the high radiation doses they had received that they could accomplish their grim task in the Special Zone.

For Vladimir Chernousenko the rectification work is anything but completed. This is why the author, in addition to his above-mentioned functions, has devoted himself to paving the way to establish an international scientific research center in Chernobyl, whose task it would be to investigate the effects of the disaster on man and nature. The first legal steps towards this goal were taken in 1991.

The author's chief motivation for writing this book is that he considers it vitally important that the world should be told the unvarnished truth about the scale and consequences of the disaster, the legacy of which will remain with us for many generations. He presents realistic estimates and new unpublished hard data from various reliable sources about the radiation pollution caused by the accident. The figures prove to be much higher than anyone dared assume up to now. We are confronted with horrendous numbers regarding the radiation pollution of the soil and aquifers in the Soviet Union. On the basis of these data, it is estimated that a territory of a least 100 000 km^2 is so polluted as to be uninhabitable. There are even estimates of an amount three times as high.

The author's greatest concern is the well-being of the people still living in this huge territory. Many of those who are still living in the polluted areas want to leave, but the problems posed by local administration and bureaucracy do not allow them to do so. For lack of precedence, the effects on their health in the long-term can only be guessed at, at the present time. But those effects are already beginning to become evident. The health statistics included in this book are a matter of serious concern and urgently call for further investigations.

Chernousenko intends to make people aware of the acute threat posed by the continuing operation of the RBMK-type reactor used in Chernobyl. There are 15 reactors of this kind still in operation in the Soviet Union – every single one of them a potential bomb. It was not so much personnel error as it was serious faults in the reactor design that was to blame for the explosion of Block 4. This is the alarming conclusion at which the author arrives after careful consideration of all factors involved.

Thus, this book is an urgent appeal to people everywhere to assist in the efforts to overcome the consequences of the catastrophe – consequences which have been gravely underestimated so far – and to do the utmost to prevent a repetition of such a disaster. The problems of Chernobyl are as pressing as they were 5 years ago when the accident happened, and they demand the attention and involvement of everyone of us.

In spite of his personal experience and his many encounters with victims of the disaster, Vladimir Chernousenko does not conclude his book by condemning everything related to nuclear energy. He makes constructive proposals with regard to the Government Rectification Program and leaves it to the reader to form his or her own opinion towards this controversial form of energy.

In the Soviet Union, lower-level officials and those who were involved in the design and operation of the faulty reactor have managed to maintain an information-ban on the events in Chernobyl (possibly even to conceal the real facts from their own government). Thus, this English edition appears before the Russian edition, which will hopefully follow suit.

In 1986, when the Chernobyl accident occurred, the attention of the press was focused on the events within the 30-km Zone around the exploded reactor. Gradually, it became increasingly clear to the people involved that the scale of the problem in terms of territory affected and of the depth of human tragedy was far larger than originally anticipated. However, now we observe a second, psychological disaster of a similar scale:[1]

For several years after the accident a true disclosure was withheld from the residents of the affected region (and from the world). Unavoidable consequences are that even correct information on radioactive contamination, on radiation doses, and on the resultant health risks is now met with disbelief, and that conflicting information is abundant. Even in regions with less radioactive contamination the people are uncertain and frightened, and even in these

---

[1] For his help with the following remarks we are grateful to Prof. A.M. Kellerer.

regions they accept grave constraints – especially with regard to food – that make normal living conditions impossible, and in many cases they want to leave their villages, even when this does not seem to be necessary from the radiological point of view. This includes many settlements and cities that could well be saved.

The misguided information policy of the past has made it almost impossible for the Soviet administration to now arrive at acceptable terms with the residents of the afflicted areas of the Soviet Union. On the governmental level, in meetings in 1989 between M.S. Gorbachev and E.R. Shevardnadze of the U.S.S.R., and Germany's H. Kohl and H.-D. Genscher, the suggestion was made that the Germans might help re-establish normal living conditions for the people in the afflicted areas. The Soviet authorities themselves felt that their information policy at various levels had caused distrust, despair, and bitterness in the affected regions to such a degree that all communication had broken down, and that an international effort was now required to re-establish some meaningful communication with the people.

On the occasion of his visit to Bonn in 1990, Mr. Gorbachev recalled these discussions. As a result, the German government promised to begin a service in the afflicted regions of the Soviet Union that would enable all concerned citizens to have their incorporated radioactivity measured once or twice a year and to thus be given reliable information on radiation doses received from consuming contaminated food. It was realized that occasional measurement campaigns for the sampling of foodstuffs, which had frequently been performed, both by Soviet authorities and by foreign groups, could not solve the existing problems. Radiation exposure due to food intake varies so much with dietary habits and with local variations in contamination that every person needs an indivudual measurement in order to judge whether he can live a normal life, whether he has to change his dietary habits, or whether he needs special help, which may include relocation.

As a result, mobile measuring units are being assembled in Germany to be brought to the afflicted areas of Byelorussia, the Ukraine, and the Russian Republic. They will be used to measure (on a large scale and repeatedly) the body activity of cesium for as many residents as possible and, thereby, to identify those – and this may well be a relatively minor fraction of the population – who need special advice and help. More importantly, however, the German staff will discuss the situation with the affected individuals. In this way it is intended to give them objective information on their environment, on the status of their health, and advice for the future. It is hoped that this will help to restore, to some extent, their faith in the future and to restore at least some some confidence in their authorities and government.

Also contradictory is the information available internationally. There are various Soviet sources (the involved ministries, institutes, and institutions at federal and republic levels) providing different and conflicting contamination data, health statics, and other measurements. The recently released studies of the International Atomic Energy Agency (IAEA) in Vienna [2] seems, at first glance, to indicate that the situation is by far less dramatic than frequently indicated; they are also in flat contradiction to the opinion voiced by V.M. Chernousenko and the sources cited by him. One part of the role of a scientific publisher is to make reliable data available to the scientific community; another important task is to keep alive the discussion on disputed topics and data in order to help to establish the truth, (independent of the pressure from or general acceptance by one or the other of the parties involved). In the case at hand, this is even more critical since the results will have serious implications on the fate of millions of people. We sincerely hope that this book will be a beneficial step towards open discussion and a clarification of the *real* post-Chernobyl situation.

---

[2] Three volumes on *The International Chernobyl Project...* (IAEA, Vienna 1991)

The Directors of the Task Force for the
'Rectification of the Consequences of the Accident
at the Chernobyl Nuclear Power Station
in the Special Zone'.
From the left: Yu. Samoilenko, V. Chernousenko, V. Golubev.

Contents

---

* An asterisk denotes chapters and sections that are more technical.

1. Black Rain

"Somewhere in the sky a chalice
of poison was spilled.
It fell like black rain
onto the smoking town."
Sinoe Sioda
a victim of
radiation sickness.

The accident which occurred at the nuclear power station in Chernobyl on April 26, 1986 is, by its scale and by the harm it caused, one of the largest catastrophes on our planet in the recorded history of mankind. It is not just a national tragedy bringing grief and misgivings to the Soviet people, but it is also an extreme ecological phenomenon on the international scale.

The territory which has been exposed to radioactive pollution is larger than 100 000 square kilometers, as stated officially in early 1990. And even now, more than 5 years after the accident – more and more patches of radioactive pollution are being discovered. The amounts of money spent on the attempts to rectify the consequences of the disaster also runs to staggering numbers: the total sum spent from 1986 to 1989 was 9.2 billion roubles (around 14 billion US $). According to preliminary estimates, more than 33 billion roubles will be needed up to 1995.

Dozens of villages and small towns have already been abandoned. But according to unofficial numbers, there are still hundreds of them where the radioactive pollution exceeds 15 Ci/km^2 – where hundreds of thousands of people still live!

Preceding page, a view of the town of Pripyat. At the horizon, Chernobyl nuclear power station can be seen (another view is shown on Plate 3 of Chapter 16).

1.1 Hiroshima and Nagasaki

Monday, August 6, 1945; 08:15 hours:
the uranium bomb is dropped.[1]
Akira Ishida, 61 years old, teacher, citizen of Hiroshima,
recalled in 1989:[2]

It was an unsettled night, the night before, three times the air raid sirens sounded, but no bombers appeared. Hiroshima was not bombed – and we rejoiced like children. Suddenly, the explosion ... I fell on to the floor and lost consciousness. When I came to my senses there was smoke all round and corpses lay on top of me. I managed to crawl out into the street – but the street had gone. No buildings, no vehicles. Everything had melted – fences, granite, roof tiles. I walked, crawled. I saw a lot of dead people. They had been burned to charcoal. I heard the groans of people whose skin had been burned off. The city had become a desert, a hot desert strewn with debris over which half-dead people were crawling. I tried leaning against an old tree. It broke like cardboard – the inside of the trunk had been burned to ashes ...

Thursday, August 8, 1945; 11:02 hours:
the plutonium bomb "Fat Man" is dropped.[3]
Susumi Nishiyama, 62 years old, writer, citizen of Nagasaki,
recalls in 1989:

I remember the sharp flash of light hitting my eyes. The shock wave shook the house. Fortunately, we were not buried. The house stood firm. It turned out that we were four kilometers from ground zero. At two o'clock I went outside. Everything that could burn was burning. Suddenly a strange rain began to fall – ashes were falling. This was the black rain. I walked as if in delirium. Nagasaki, our beautiful city, which we had compared to Naples, was crushed. It was impossible to count the bodies in the river ...

---

[1] The bomb was 3 meters long with a diameter of 0.7 meters and a weight of 4 tons.

[2] *Author:* Following a request of mine, the correspondent of "Komsomolskaya Pravda" was kind enough to conduct this interview and the following one in Northern Korea in August 1989.

[3] This bomb had a length of 3.2 meters, a diameter of 1.5 meters, and a weight of 4.5 tons; it was meant for the city of Kokure, but that city was covered by fog, so the American bomber diverted to Nagasaki.

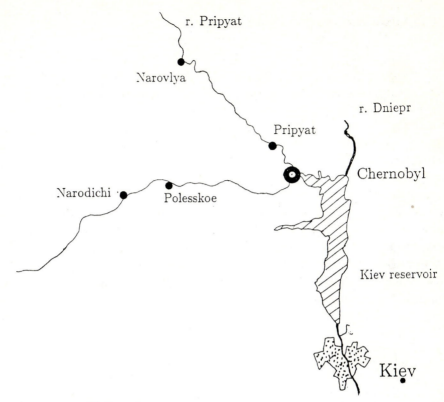

Fig. 1.1. Sketch of Chernobyl and its surroundings

1.2 Chernobyl [4]

Saturday, April 26, 1986; 01:23:40 hours:
Explosion at Block 4 of the Chernobyl nuclear power station.[5]
Vladimir Starovoitov,[6] 31 years old, concrete worker, a resident of the
village Burakova close to Pripyat, remembers:

---

[4] Chapters 8 and 10 specifically address some of the problems alluded to in this section.
[5] Operating with an RBMK reactor. RBMK stands for "reaktor bolshoi moshchnosti kanalnyi" [high-power channel reactor]. The reactor is housed in a concrete shaft measuring about 26.6 × 21.6 × 25.5 meters; its fuel load is about 192 tons of uranium. For further details on the RBMK reactor see Chapter 3 and Appendix B.
[6] Starovoitov participated in the construction of the third phase of the Chernobyl Atomic Power Station (Blocks 5 and 6). This interview with the author was recorded on April 26, 1990, in Isolation Unit No. 2 of the Radiation Pathology Department in Kiev.

The night was warm, this was on the eve of the holidays and some of us wanted to go fishing at the pond, the cooling pond of the Chernobyl station. We always fished there because it is warm all year round and the fishing is always good.

I was near Block 4, about 500 meters away, when I suddenly heard a loud clap. Then came something like the sound of an explosion. I thought it was the steam valve, which we used to hear from time to time. Then in a couple of seconds a bright, blue flash was followed by an enormous explosion. When I looked at the Block, I saw that there were only two walls of it left. The roof and two other walls had been destroyed. The structure was in ruins, water was pouring out, bitumen was burning on the roof of Block 3.

One or two minutes afterwards it started to rain. I thought at first that it was normal rain, but when I came into a nearby building and looked in a mirror I saw my face was covered with soot. Then I realized that I had been out in radioactive rain.

There was no alarm siren.

I took off my shirt and washed myself. In about 20 minutes I felt very bad. By then, fire engines were already at the site and I could see firemen on the roofs. I phoned to the controller's office and told them that I was unwell, upon which I heard: "You are in the radioactive fallout zone, run for it!" I told them that I simply couldn't move – my legs wouldn't carry me, and then I fainted.

I came to myself in Medical Center No.126 in Pripyat. I was sick nonstop. My temperature reached 40°C. They put me on a drip and used five bags of solution on me. Then I felt much better. Then they started bringing lads who had been on duty on that night and firemen. It was as if a war had started.

The first batch of lads from the station was taken by air to Moscow on April 26 at 21:00. I was taken there in the second group, which went by bus the next day, April 27 at 12 noon to the Borispol airport in Kiev. And from there by a special flight to Moscow, to Clinic No.6.

There were two to three of us in each ward. At first, our condition was not too bad, we even visited each other there. Then came the crisis.

After May 9 the lads started to die. It was horrible. Then they moved all of us into individual wards and told us not to walk. We didn't feel like it anyway. There was an awful weakness and dizziness. Our hair was already falling out. Terrible internal pains began. I just wished everything would come to an end.

First one of the firemen died; then some lads from the station.

They didn't tell us who had died. But we were well informed by the soldiers on duty in the hospital – this was a restricted hospital belonging to the Ministry of Engineering [7] and the soldiers had been brought in to carry out decontamination in the corridors and wards. They told us that when we were brought in we were dirty and were glowing.

---

[7] MinSredMash – Ministerstvo Srednego Mashinostroyeniya

The soldiers told us later that they had seen (photographic) slides in that hospital of burnt sailors from atomic submarines and of workers from nuclear stations where explosions had happened before. Those lads had had doses of 10 to 15 Gy,[8] and they all died later. As for us, each was waiting for his turn, too... I was in the hospital until September 15, 1986.

No one told me the dose that I had received. And when I pressed with my query there in Moscow they told me in strict confidence that I had had 3.8 Gy. Then, later in Kiev I had a medical examination at a medical commission,[9] then I saw by chance the figure of 5.4 Gy. Later, one night I sneaked into the intern's office and got out my medical record card with the results of the chromosome analysis, and there I found my true dose – 6.8 Gy. Somebody told me that this is a lethal dose... I want to live... I am still very young...

Now I am a group 3 invalid. I am constantly ill. In 1988 I had plastic surgery on my legs, but the burns and ulcers on my legs don't leave me in peace. My muscles are in agonizing pain – as if somebody is squeezing them in a vice; my bones are aching and clicking. When the weather changes I have diabolical pains all over my body. I am losing my eyesight.

I am admitted to hospital five to six times a year. I suffer from weakness. I know the doctors are doing their best, but they haven't got the right medications. They are fantastic people.

I heard that there are special treatments and equipment in the West. They did offer them to us, but our government declined the offer. We all need some help – medical, social, and financial. The lads I was with in the hospital have meager pensions which don't provide a living for a family.

In the days of the disaster they were called heroes. They had extinguished the fire, they had been on the roofs when it was over 1000 Roentgen/hour,[10] they had taken part in the construction of the "Sarcophagus", they were saving the country. Today nobody remembers them. They are ill and dying.

The Chernobyl staff made a list of the lads who died. There are more than 100 of them. And all those, thousands, who were in the Special Zone [11] in 1986–87 are probably dead by now. And there are plenty left alive as invalids, like me.

Just look at Burakova, which is in the 10-km Zone. It was evacuated on May 4, 1986, first to the Tolstyi Les,[12] then to the Makarovskii region. All these people – children and adults – are ill now.

---

[8] The energy dose gives the total absorbed radiation energy per mass unit. It is measured in Gray [Gy]; up to the end of 1985 the unit Rad [rd] or [rad] was used. 1 Gy = 100 rd. For further units and their relation to each other, please see Appendix A.

[9] VTEK – vrachebnaya trudovaya ekspertnaya komissiya

[10] 1000 R/h, in the scientific notation. This quantity gives the ion dose (of the external irradiation). Since 1985, Coulomb per kg is the official unit, i.e., [C/kg]. For further details see Appendix A.

[11] 10-km Zone around the power station.

[12] In May 1986, radiation levels there were 10 to 20 R/h.

I live with my mother, born in 1923, and she is also very ill. We have a vegetable plot which feeds us. I've got a brother and two sisters. At the time of the accident my brother was in the army. My sister, born in 1957, was working at the concrete works by the Chernobyl station on the night of the accident. They were working on the construction site of phase 3. No one warned them about the accident, and they finished their night shift not knowing what had happened. At present she lives in the Volodarskii region. She suffers from liver and gall bladder trouble and she has a heart condition and headaches. Both of her daughters – one born before and the other after the accident – are very ill. My sister asks me to get medicine for her, but you can't get it anywhere, except on the black market, but the prices are well beyond the reach of my pension.

My elder sister also lived in the Special Zone in the village of Chisto-golovka. They were also evacuated after May 4, 1986. Her younger daughter, who was 2 at the time of the accident, spent days playing in the sand box in the garden. Her 10-year-old son went to school as usual in Pripyat on April 26. No one warned them about the accident or the danger. It was only later in the day, when tanks and soldiers in gas masks came to Pripyat, that people found out that the Block was burning. The boys and the adults from this school ran to the bridge to watch the fire – it was only 1 kilometer from Block 4. Today both my niece and nephew are sick – they have eczema on their hands and legs, they suffer from liver and stomach disorders, they're anemic, and they have headaches.

Since the time of the accident they have not had a medical check-up. There was no iodine treatment on April 26 or 27. Later, in Kiev, they had blood samples taken, but they were not given the results. My sister spends all her time searching for medication for them.

One of my friends, Yanov, who was on duty at the control check-up point by the station on April 26 – 27, 1986, is a patient in Section No.1 at the hospital. He took part in the evacuation of Pripyat and was responsible for all the buses which were used in the evacuation. He was exposed to quite a bit of radiation there. The radiation control staff said later that the levels there had been 200–400 R/h. He had first category radiation sickness. His two daughters have been born since the accident, both are very sick; they spend all their money on medicine.

Academician Il'yin [13] claims – and the newspapers say so, too – that there is not a single case of radiation sickness among civilians. But why then are there so many people like me? Why is it so difficult to get hospital treatment?

I am in a better position, because I was among those first 300 who were diagnosed as suffering from radiation sickness, while the rest didn't get this

---

[13] Director of the Institute of Biophysics in Moscow.

diagnosis. Mrs. Guskova [14] came from Moscow on several occasions. She came together with the commission which rejected the diagnosis of radiation sickness which had been made before. The result is that the lads can't get proper treatment in the hospital. Recently, I heard they were going to close the Department of Radiation Pathology in Kiev. Well, if that happens what will become of us?

Fig. 1.2. A medical check-up at the Chernobyl nuclear power station. (Photo taken June 5, 1986.)

---

[14] Chief doctor of the department of Clinic No.6 in Moscow which has a long-standing experience in dealing with radiation victims. The first victims of the Chernobyl disaster were also flown to Moscow to be treated in this hospital.

1.3 On the Design of the RBMK Blocks [15]

Five years have gone by.

Thousands of articles have been written and hundreds of books have been published on different aspects of the catastrophe of Chernobyl, a catastrophe whose consequences concern everybody. In the light of more recent data it appears that – to varying degrees – practically all areas of the earth are affected.

As previously, the remaining 15 RBMK reactors (at the locations shown below) which continue active service are reasons for grave concern.

Several independent investigations into the reasons for the catastrophe of Block 4 of the Chernobyl nuclear power station revealed not less than 32 infringements of nuclear safety committed during the design and construction of the active zone (core) and of the control and safety systems of the RBMK reactor.

However, the monopoly and secrecy which were always related to nuclear installations within the Soviet Union ensured that such projects were beyond any criticism. Thus, the deficiencies in the construction of the reactors were covered up and not eliminated.

Fig. 1.3. Locations of the RBMK reactors still in operation

---

[15] Further details on this topic are given in Chapters 2, 3, and in Appendix B.

A more official view on "The Nuclear Accident in Block 4 of the Chernobyl Nuclear Power Station and the Safety of the RBMK Reactors" give the following excerpts from an unpublished report by A.A. Yadrikhinskii, Nuclear Safety Inspection Engineer of the USSR State Atomic Energy Survey Commission (Kurchatov town, RSFSR February, 1988):

The nuclear accident on Block 4 of the Chernobyl Station... was the result both of blatant safety infringements in the design of the core and of the control and protection mechanisms of the reactor committed by the scientific advisor and Chief Designer and of their failure to adhere to the guidelines laid down by the State Atomic Survey Commission.

The close examination of the RBMK design, carried out at the Kursk nuclear power station, found 32 infringements of Safety Regulations.

Radiation emission was no less than 80% of the core (with a total of 192 tons), which amounted to 6.4×10^9 Ci.[16] If we divide the figure by the population of the whole earth (4.6×10^9 people) then we get 1 Ci per person.[17]

The radiation levels of the emissions from the Chernobyl disaster exceed 16 to 27 times the maximum figure estimated as resulting from a hypothetical accident, in which the fuel rods melt down and the safety mechanisms are destroyed – this maximum figure was calculated as 3–5% of the core content.

It is practically impossible to eliminate all radioactive substances from the subsoil and soil in the contaminated area. It is also not reasonable to hope for natural decay of the radiation. The radiation levels given off by the substances emitted from the reactor will in the first 100 years decrease 5 times from 5×10^{12} to 1×10^{12} and, in 1000 years, 1000 times to 1×10^9.

One way of illustrating the danger is to calculate the volume of water required to dilute the radioactive material to the maximum permissible concentration. The 15 m^3 of radioactive substances emitted from Block 4 at this time could be diluted in $15 \times 5 \times 10^{12} = 75\ 000$ km^3 of water. In 100 years, 15 000 km^3 would be needed. In 1000 years 15 km^3 would be the required amount. For comparison: the total outflow of the world's rivers is 36 380 km^3 – i.e., its use to dilute the radiation would take 50 years before the radioactive emission from Chernobyl will be brought down to the permissible level.

Disasters on the scale of the Chernobyl accident lead to harmful effects on the population, territorial losses without any military action, and to thousands of billions of roubles'[18] worth of damage, and are, therefore, hard to justify by the need for electric power.

---

[16] The old unit for the activity was the Curie [Ci] which has been replaced by the Becquerel [Bq]; 1 Bq = 1/s implying one decay per second. 1 Ci = 3.7×10^{10} Bq and 1 Bq $\simeq 2.7 \times 10^{-11}$ Ci $\simeq$ 27 pCi. See also the Appendix.

[17] Naturally, the implications are not that everybody received such a dose, but such crude numbers certainly help to illustrate the scale of the accident.

[18] At that time 1 rouble was roughly 1.5 $.

It was the secrecy and lack of accountability of our nuclear science, and its refusal to open itself up to discussion and criticism which made it possible for dangerous design faults to lead finally to a nuclear accident of this scale. No technical design plan of any one of the existing nuclear power stations in the USSR is available. The Soviet nuclear industry presents its projects as works of near-genius so that they apparently feel that reactor design deficiencies and infringements of safety regulations have to be hushed up to go unnoticed and – what is significantly worse – uncorrected for years and even decades. Economical reactor operation is pursued at a definite cost in terms of nuclear safety.

We should not tolerate a situation in which scientific leaders and scientific organizations, any more than any other people, are above the law or the safety regulations or not open to the scrutiny of the State. This should be independent of how many they are or how exalted the position they occupy, no matter what high honors they have received. State control of safety should be exactly that, i.e., State control answering only to the legislative bodies. (Within the USSR, to the Supreme Soviet of the USSR and not to the Technical and Operational Control Bureau of the Council of Ministers, as at present.)

Our plight as a nation and our disgrace in the eyes of the world arising from the state of our nuclear power industry should compel us to carry out a truly detailed, objective, independent, official enquiry into the Chernobyl accident.

RBMK Reactors: A significant fraction of the nuclear energy capacity of the USSR, ~50%, or 16 million kW, is produced by RBMK reactors operating at the following nuclear power stations (NPS):

- 4 000 MW Leningrad NPS with 4 blocks;
- 4 000 MW Kursk NPS with 4 blocks;
- 3 000 MW Chernobyl NPS with 3 blocks (formerly 4);
- 2 000 MW Smolensk NPS with 2 blocks;
- 3 000 MW Ignalin NPS with 2 blocks;

("M" denotes "Mega" implying a factor of 10^6). All these blocks are powered by a reactor that is a descendant of the industrial uranium-graphite reactors of the Siberian NPS-type which could deliver 0.6 million kW.

The scientific directorship of the RBMK project is based at the Kurchatov Institute of Atomic Energy (IAE), the main designer is the Scientific Research and Development Institute of Electro-Technology (SRDIET; in Russian: Nauchno-issledovatel'skii institut energotekhniki – NIKIET), and the general planning and construction work was with the S. Ya. Zhuk institute "Gidroproyekt". The proponents of the project were able to convince the Government of the complete nuclear safety of the RBMK; in spite of doubts about their nuclear safety.

1.4 The Ukraine [19]

After 1986, people in the most strongly affected areas of the Soviet Union started to divide their lives into two periods: the period
before Chernobyl and the one after Chernobyl. Sometimes people
said: "That was before the war", the implication being that 1986
was similar to wartime, in terms of danger to health and well-
being. It also reflects the panic which was experienced during the
evacuation of several larger cities which were subject to radioactive pollution or when the children were sent away to safer parts
of the country.

A medical investigation of the public health in the provinces of
Kiev, Zhitomir, and Chernigov in the Ukraine, conducted in 1989,
indicated that the health of very second resident is damaged.

Large numbers of militia were on duty in Chernobyl immediately after the disaster, taking an active part in the evacuation
of the population from the 30-km Zone and guarding the deserted
villages and towns. They had to stay for rather long periods in regions with dangerous radiation levels. Thus, it is no surprise that
many of them suffer now from the consequences of their extended
stay in the danger zone.

Fig. 1.4. A sketch of the Ukraine

---

[19] For more data and discussions see Chapters 8, 10, 11, and 12.

The Health Commission: *Unpublished data collected by the Ukrainian Ministry of Health concerning health checks on the population of the provinces of Kiev, Zhitomir, and Chernigov, as taken during 1988 and assembled in January 1989. Under public pressure, the Ministry of Health of the USSR created (about 2 years after the accident) an All-Union Register meant to monitor medical data and statistics on the health of the population in the area affected by the accident.*

Table 1.1. The number of people included into the recently created All-Union Register of these three provinces of the Ukraine (according to the Ukrainian Ministry of Health).

No.	Province	Individuals examined	Adults	Children
1	Kiev	108 498	89 322	19 176
2	Zhitomir	26 833	21 273	5 560
3	Chernigov	52 412	25 165	27 247
4	Total	187 743	135 760	51 983

Table 1.2. Breakdown of the ailments affecting the adult population of Kiev, Zhitomir and Chernigov provinces in 1988. It should be noted that the number of people investigated within this series of check-ups (also by the Ukrainian Ministry of Health) is larger than the number of people entered into the All-Union Register referred to in Table 1.1.

No.	Ailment	Number of cases	
		Kiev	Chernigov and Zhitomir
1.	Neoplasmic growths	1 445	2 005
2.	Diseases of the endocrine system	2 442	3 046
3.	Psychological disorders	2 535	3 051
4.	Diseases of the nervous system	5 116	7 037
5.	Blood conditions	25 589	36 065
6.	Diseases of the respiratory tract	17 089	21 048
7.	Diseases of the digestive tract	8 799	11 863
8.	Diseases of the skeleto-muscular system	5 998	9 152
9.	Hyperplasia of the thyroid gland	7 230	7 220
	Altogether this yields	76 243	100 487

"Moscow News" No.31, July 30, 1989. *An excerpt from information supplied by the office of Internal Affairs of the Kiev province Executive Committee. It refers to the state of health of the employees of this organization, i.e., to the militia men who were on guard in Chernobyl during the time after the accident:*

57 employees are suffering from radiation sickness, 4750 from "vegetative (blood vessel) distonia",[20] in 1500 cases existing chronic conditions, including respiratory diseases, diseases of the heart and blood vessels, diseases of the stomach and intestinal tract have grown more acute...

Fig. 1.5. Chernobyl, robots are prepared to get ready for work. (Photo taken June, 1986.)

---

[20] The Russian term coined for this is "vegetososudistaya distoniya", describing an effect of radiation in which the blood vessels spasm and contract, thus affecting the cardiac rythm and leading to a general state of sleepiness and inactivity.

1.5 Politics Versus Rectification [21]

In the course of time, with the accumulation of more and more bits of new information about the scale of the catastrophe, practically all people in the Soviet Union have started to understand the situation. The concentration of efforts on the (30-km) Zone was not so much related to objective necessity, but rather to political motivations. There was no need to bring the other blocks of the Chernobyl nuclear power station back on line – their contribution to the energy supply of the Soviet Union was not so significant. Instead of these efforts directed towards a political demonstration, it should have been considered necessary to save the people living in the huge region polluted by the fallout of the explosion, with its radioactive isotopes ranging from cesium to plutonium.

Vasilii Omelchenko, 41 years old, graduate of the Kiev Polytechnic Institute. Since May 1986, he has participated in the rectification work in Chernobyl. He helped to set up the mobile factories which produce decontaminating fluid. He has worked for the Ukrainian State Supply Agency, organizing material and technical supplies for the restricted zone. For length of continuous service at Chernobyl and for active participation in the emergency program he has no equals in the zone.[22]

C: What is it that prompts you, a person who has reached a fairly eminent position, to come forward with this disturbing information about what went on inside and outside the Zone?

O: I will only talk about what I personally know and have actually encountered. Only one thing prompts me to share this information – the hope that this information will enable people to make sensible decisions as to what should be done in the 30-km Zone, and not only the Zone, but about all the work of containment and rectification at Chernobyl.

C: You say, "information which will help people to work out sensible solutions". What sort of solutions were adopted during the first 5 years after the accident, if there still is a need for information on which to base sensible solutions?

---

[21] For further details see Chapters 4 and 5.

[22] The author (C) conducted this interview with V. Omelchenko (O) in Chernobyl in February 1989.

O: As far as I can see, the emergency measures taken at the very beginning were of a purely political nature. They served political purposes. No one thought about the economic consequences. No one even considered the loss that the State would have to bear if these political aims were to be achieved.

More specifically, I consider that it was wrong to try to get the station working again in so short a time and in such a hurry as there was in 1986 and the years that followed. The loss to the State was not just that of 9.5 billion roubles that was officially announced, but considerably greater. I have in mind, especially, the damage done to the health of the personnel who worked in the emergency teams. That was staggering.

C: What were the particular political considerations which prompted all this?

O: I think that the image they tried to present to the world was like this: that an accident on such a scale could not happen here; that the state could cope fairly easily with the consequences of such an accident. That was the attitude at the beginning – although they knew from the very onset that the cost would be colossal. They took this political decision to deny that here we had a disaster of global proportions – to show that the state could cope with it and, in doing so, demonstrate the stability of the existing political system.

But they should have concentrated on saving the people, not the station. The resources that were poured into the rehabilitation of the power station and not the work done in the Zone, which proved later to have been completely ineffective, and first and foremost the so-called structural deactivation.

C: Meaning?

O: The use of heavy equipment to pile all the most highly radioactive debris into so-called temporary storage. In other words, digging a pit and burying it all. There was no urgent need for such measures. They should have dealt with the areas which were known to have suffered. Take Narodichi, for example. All the experts knew about that even in May, 1986, and nobody should try to conceal the fact. The direction the plume (the cloud of emissions from the ruined Block 4) and the radiation levels in the plume were known back in May, 1986. To say that the matter had been insufficiently studied was not exactly true. Possibly, more research work was needed, but the direction and the dimensions of the radioactive plume were well known. We should have put all our resources in 1986 into those places which had been subjected to high levels of radioactive contamination and where people were living. We should have fed them with the right foods and dealt with the social problems. The station could have waited. There was absolutely no urgent need to get it working again. Even the Ukrainian energy minister, speaking at a meeting of

"Green Light"[23] held by People's Deputy Yu. Shcherbak, confirmed that the Chernobyl nuclear power station played no significant part at that time in the energy budget of the Ukraine.

The classic example of absurd misuse of material and technical resources is the bombing of the ruined reactor with lead. This is the most perfect, classic example of the way that people, instead of basing an operation on laboratory experiment, turn the whole operation into an experiment on a gigantic scale. According to the documents, they brought 6000 tons of lead into the Zone. There are various estimates of the amount that was thrown into the reactor, but I reckon that they dropped about 3000 tons (because after the deactivation we put about 3000 tons back into the national economy). The result? Even before it hit the reactor the lead evaporated – causing lead poisoning among the people and lead contamination of the area.

C: What was the cost of the work done in the Zone?

O: From the very beginning nobody counted the cost in human life. And why bother counting? The decision had been taken to rehabilitate the station, to get it working, to show that under our social system everything is possible. And they proved it. They threw people on the fire. If they had needed to burn more people, they would have done it.

The saddest part of the emergency work was this: the Ukrainian government should have been involved from the beginning; all the resources should have been directed to the areas that had been affected by the accident, instead of being used to get the station started up again.

C: What should be done?

O: There should be a decision right now to close down the Chernobyl station, because while it carries on operating there will be a continuing need to keep the people who are working there in place. Maybe even more staff will be needed. Secondly, there should be a scientific observation program in the Zone, involving a minimal number of people in the research work.

---

[23] In the Ukraine, this name refers to the unofficial organization of the "Greens", an evironmental group.

Fig. 1.6. Looking from a helikopter at the progress made in building the Sarcophagus. ▶ (Photo taken in September 1986).

1.6 Byelorussia [24]

The three Soviet Republics which had to bear the most serious onslaught of radioactive nuclides are Byelorussia, the Ukraine, and the Russian Soviet Republic. It is estimated that within the territory of the latter more than 70% of the emitted cesium-137, strontium-90, and plutonium are to be found. Investigations of Byelorussian scientists showed that almost the entire territory of the republic is polluted by radionuclides. Even the most northern regions of the Vitebsk region seem to be affected.

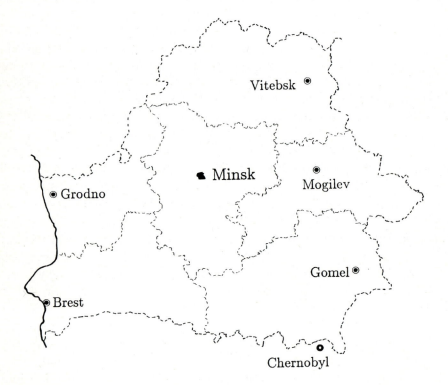

Fig. 1.7. The six provinces of Byelorussia and their capitals are shown

---

[24] For further details see Chapters 11 and 12.

The Byelorussian Academy of Sciences *prepared a report on studies connected with the emergency work following the Chernobyl accident from which the following excerpts (from the subdivisions 1 to 4) have been taken. The report was signed by the President of the Academy of Sciences of the BSSR, Academician B.P. Platonov, and also by the Director of the Institute of Radiobiology of the Academy of Sciences of the BSSR, Corresponding Member of the Academy of Sciences of the BSSR, E.F. Konoplya. March 1989, Minsk, Byelorussia.*

1.   Radiation and the Environment

The areas most seriously contaminated as a result of the Chernobyl accident were in the provinces of Gomel and Mogilev. Gomel province is now contaminated with long-living radionuclides: cesium-137, strontium-90, and plutonium – the last two found mainly in the evacuated zone and the areas adjacent to it – and to a lesser extent, cesium-134, ruthenium-106, and cerium-144. Cesium-137/134 and, to a lesser extent, ruthenium-106 are also found in Mogilev province. They have also been discovered in the provinces of Brest and Minsk. A level of cesium-137 contamination in the soil that exceeded 15 Ci/km^2 was recorded [25] in the autumn of 1988 in 408 of the republic's towns and villages – i.e., in areas inhabited by more than 100 000 people. In a 624 km^2 area of Gomel and Mogilev provinces which contain towns and villages and a population of 12 000 there is contamination of more than 40 Ci/km^2. The total area of Gomel province where contamination with cesium-137 is in excess of 5 Ci/km^2 is 9 332 km^2 (23.1%). The corresponding figure for Mogilev province is 5 376 km^2 (18.5%), for Brest 120 km^2, and for Minsk 60 km^2. Areas where contamination reaches 5 Ci/km^2 have been found in Grodno province and levels of up to 1 Ci/km^2 have been recorded in Vitebsk province. In the whole republic 310 000 people are living in 1052 towns and villages where the contamination level is 5 Ci/km^2 or higher.

Studies of the extent and character of strontium-90 contamination in Byelorussia show that the bulk of it lies in the evacuation zone, where concentrations exceeding the maximum permissible one are to be found in the surface level of the soil. There are particular spots where the strontium-90 content is greater than 3 Ci/km^2, especially in the south of the Narovlya region. Levels of $1-3$ Ci/km^2 have been found in the regions around the towns of Khoiniki, Bragin, Narovlya, around the settlement of Buda-Koshelevo and other places. Otherwise the strontium-90 content is below maximum permitted levels.

---

[25] A crude estimate based on the assumption that the radiation is due to cesium-137 (different nuclei yield different radiations) implies that a person living in an environment with 15 Ci/km^2 would receive within 1 year the dose of 2 rem of external radiation. International norms consider 0.1 rem within 1 year to be a permissible dose for the normal population. (For more details, see the Appendix.)

In estimating the contamination of the republic's territory with transuranic elements (plutonium-239/240, americium-241/242), it is necessary to point out that the bulk of these fell within the 30-km Zone.

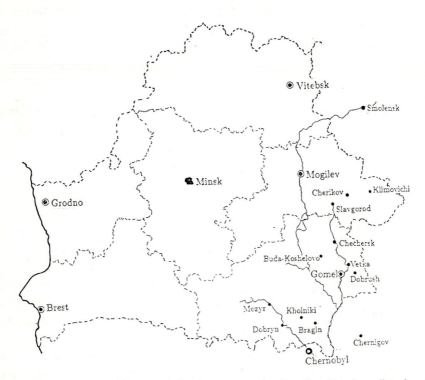

Fig. 1.8. The eastern part of Byelorussia has been most seriously polluted by the radioactive cloud from the Chernobyl disaster

On the edge of this zone the concentration of plutonium isotopes is at the maximum permitted level (0.1 Ci/km^2 in the soil and 3.0×10^{-14} Ci/km^2 in the air at ground level). Experiments show that when the wind blows from the south 24-hour average radiation levels can reach the permissible maximum at a distance of 100 km from Chernobyl. Measurements were taken while agricultural work was carried out in the Bragin, Khoiniki, and Narovlya regions in the autumn and spring of 1987–88. When the dust content of the air exceeded 0.5 mg/m^3, concentrations of plutonium-239/240 were two to five times greater than the permitted maximum. When stoves burning local organic fuels were used for heating during the autumn-winter period in these same areas there could also be a rise, at particular times and places, of the level of plutonium in the air.

In the rest of the republic plutonium-239/240 levels are generally 1.5 to 2 times higher than the background level while remaining below the permitted maximum. Up to now, increased amounts of transuranic elements have been found in several localities (in the northwest area of the town of Pripyat, to the northwest of the village of Lomachi in the Khoiniki region, to the east of the town of Dobrush, etc.).

When estimating levels of plutonium contamination it must not be forgotten that it forms a persistent finely-dispersed aerosol and that the way it usually penetrates into the organism is via the respiratory tract. Maximum plutonium concentrations in the air occur in the spring and autumn when work is being carried out in the fields. 24-hour average levels of airborne plutonium can rise as a result of this activity, not only in the towns and settlements on the edge of the evacuation zone, but also at considerable distances from it.

Hence, the plutonium content in the air sometimes reaches

5×10^{-4} Bq/m^3 in Mozyr,

2×10^{-4} Bq/m^3 in Gomel,

9×10^{-5} Bq/m^3 in Brest, and

1×10^{-5} Bq/m^3 in Minsk.

The pre-accident background level in the air was $2 - 5 \times 10^{-8}$ Bq/m^3.[26]

The maximum permissible concentration in the air is 1.1×10^{-3} Bq/m^3.

Contamination studies in the Byelorussian republic have shown that the danger of this type of contamination is not only to be measured in terms of the amount of radioactive isotopes deposited on this or that territory, but also depends substantially on their chemical nature and on the forms in which they are found in the upper layers of the soil. This last will, to a considerable degree, determine the character of the interaction between the radionuclides and the constituents of the soil and also, in consequence, their mobility and redistribution, i.e., the way they move in the soil cover, in mud and surface water, from soil to plant and, therefore, in the food and biological chains.

A survey of radioactive contamination of plant life has shown that it is very unevenly distributed but, on the whole, it correlates with levels of soil contamination.

Research carried out in 1986 showed that practically the whole territory of the republic was contaminated with radionuclides, the sole exception being the northern part of Vitebsk province.

The contamination of the republic is very patchy. Even in the most heavily contaminated zones at Narovlya, Khoiniki, and Bragin there are places where radiation levels can be 20 to 30 times higher or lower than other places only 0.5 to 1.0 km away.

---

[26] These units denote the activity, 1 Bq stands for one decay per second. For further information, please see Appendix A.

The heaviest contamination was suffered by the forests, which acted as natural filters. The places without forest cover – meadows and arable land – are, as a rule, three to five times less contaminated. In 1988, the radiation levels in the natural phytocenoses stabilized and are now changing very slowly, because slow-decaying isotopes are involved.

Persistent radioactive contamination of natural phytocenoses affects territory equal to one-fifth of the total area of the BSSR. This territory lies in the southeast of the republic along the line Stolin – Loban – Shatilki – Buda-Koshelevo – Slavgorod – Cherikivo – Klimovichi. In the province of Minsk a large patch of radioactive contamination has been discovered in the Volozhinsk region. In the territories mentioned there should be no picking of mushrooms or berries or picking of plants for medical purposes.

It has been discovered that there are high levels of radioactive contamination exceeding $50 - 100$ Ci/km^2 in natural phytocenoses far beyond the limits of the evacuation zone – in the regions of Dobrush, Vetka, Chechersk, Cherikov, in the provinces of Gomel and Mogilev.

A peculiarity of the contamination of natural phytocenoses with radioactive debris is that the radionuclides are still concentrated in the very top layer of the soil, i.e., to a depth of 5 cm. Therefore, plants of natural phytocenoses contain a large quantity of radionuclides, even when the soil is not badly contaminated.

The nuclide content of subsurface water depends very much on the amount of surface contamination. The greatest concentration of strontium-90 has been discovered in wells in the Nizhe-Pripyat zone, within the region that would under normal operation of the Chernobyl power station still "feel" its presence. For example, a well in the village of Aleksandrovka in the Narovlya region contained a strontium-90 concentration of 1.1×10^{-9} Ci/liter.[27] The cesium-137 content of wells in the villages of Sentsy and Drozhki in the Khoiniki region was 6.1×10^{-9} Ci/liter. Basically, the concentration of radionuclides in the subsurface waters of this area vary from 1.2×10^{-11} to 8.0×10^{-11} Ci/liter for strontium-90 and 1.0×10^{-10} Ci/liter for cesium-137.

On the basis of the research that has been carried out, it is possible to surmise that in the next few years concentrations of radionuclides in the underground waters of the territory in the region of influence of the Chernobyl power station operating under normal conditions will not change substantially from 1988 levels. Local concentrations of radionuclides may occur in surface water and drainage channels, depending on the overall level of contamination, on the actual geochemical conditions in particular localities, and on the nature of agricultural activity in the catchment areas. In the particularly unfavorable situation that exists in the Sozh and Nizhe-Pripyat areas which are within the

---

[27] This quantity describes the activity per liter of the liquid, i.e., its degree of radiation. For the degrees of radiation officially considered to be temporarily permissible, see Table 11.3 and also the Appendix.

reach of the normal action (e.g., pollution) of the Chernobyl power station, the cesium-137 content of the waters of the rivers Pripyat and Sozh can reach values as high as, for example, in the case of the river Pripyat, 2 to 20×10^{-10} Ci/liter. Radionuclides build up in the beds of watercourses and can be a long-lasting source of contamination in water systems, also affecting fish. This must be borne in mind by farmers and others.

A forecast of the distribution of radionuclides in the soil cover was based on mathematical processing of the research data by means of a model which describes with reasonable accuracy all the mechanisms by which the radionuclides are moved from place to place in natural conditions. The forecast showed that the situation in the republic with regard to radiation will not change substantially in the next few years and will remain complex. Natural decontamination of the soil by vertical migration will be extremely slow: despite the fact that a significant proportion of the radionuclides are in mobile form, most of them in the coming decades will remain in the upper layers which are affected by ploughing. The reduction in the number of radionuclides in the topsoil will basically be brought about by their natural decay, by secondary wind dispersal, and by horizontal and vertical migration, as well as by the redistribution of contamination to clean areas by the movement of agricultural products, machines, and wild animals.

It is necessary to point out that, because of its low mobility, the amount of plutonium in the topsoil will remain roughly the same in the coming years. Second, in 10–15 years' time a noticeable contribution to the total dose from transuranic elements will be made by americium-241, produced by the transformation of plutonium-241. Americium-241 is highly toxic, more mobile than plutonium and has a half-life of 433 years. Third, the slow natural removal of contaminants from the topsoil will imply an increased concentration of them in the air over a long period. The proved relationship between the intensity of contamination and the state of the flora and fauna predetermines the forecast: it will be in accordance with the dynamics of the radiation situation as described for the coming years.

In the assessment of the ecological consequences of radioactive contamination it is essential to take account of the fact that a certain proportion of the radionuclides, especially in the southern regions of the republic, occur in the form of "hot" particles. The area affected by hot particles, the particle concentration, the kind of particles that are dispersed and their vertical migration in the soil are directly determined by their concentration in the near ground-level air and, in turn, determine the extent to which they are able to penetrate into the human organism, particularly the lungs. This last fact must be remembered when the question of residence and economic activity in the contaminated areas is considered.

2. *Biological and Medical Aspects*

Post-mortem examinations of people who died in the provinces of Gomel, Mogilev, and Vitebsk in 1986–88 showed the presence in their tissues of radionuclides of various activity spectra. The greatest accumulations of the isotopes of cesium and ruthenium were found in the muscles and the spleen, strontium accumulated in the bones, and plutonium in the lungs, liver, and kidneys. It is important to note that the preliminary data gathered so far do not indicate a clear correlation between the accumulation of radionuclides in the organs and tissues, on the one hand, and the nature of the environmental contamination of these regions on the other. In fact, the deposits of radionuclides in the organs and tissues of people from various regions of Gomel province are roughly identical.

In the province of Vitebsk, although smaller accumulations of radionuclides in the organism are involved, they did not correspond to the degree of contamination of the regions of this province. One reason for this may lie in the quality of food products brought into the area. Along with this, the migration and mobility of the radionuclides may be such that, even in places with a low level of contamination, agricultural products may be affected and might not be checked before consumption. Finally, the attainment of a steady level of accumulation may possibly depend on the existing balance between consumption and expulsion of radionuclides from the organism.

Studies of the main physiological systems showed consistent alterations in function. Lymphopenia was discovered in 10–40% of people from several regions of Gomel and Mogilev provinces. A rise in the incidence of anemia and myelotic dysplasia was noted. In 35 – 67% of individuals of various age groups studied, changes in the immune system were discovered. Not only was there evidence of a lowering of the level of immunity, but also of the emergence of the requisite conditions for the occurrence of auto-immune pathology. Also noted were disturbances of the regulation of the hormonal function of the thyroid gland and a higher frequency of thyroid hyperplasia, as well as changes in the action of the hormones in the organs and tissues.

Substantial changes were discovered in the genetic apparatus of the lymphocytes of the peripheral blood and the bone marrow in adults and children. In women of Gomel and Mogilev provinces, on the average, the occurrence of dicentrics and rings in the lymphocytes of the blood, the most characteristic indicator in biological dosimetry was, respectively, $0.8 \pm 0.04\%$ and $0.10 \pm 0.05\%$ among women who had recently given birth, and $0.15 \pm 0.04\%$ among pregnant women of Gomel province, which is significantly above the level of the control group with $0.04 \pm 0.03\%$.

Observation of the dynamics of the cytogenetic effect [28] among new mothers and newborn infants from these provinces, who were studied in 1988, revealed a

---

[28] That is, genetic aberrations on the level of individual cells.

definite increase in the total number of chromosomal mutations, predominantly in the form of chromosome aberrations with a $20 - 100\%$ increase in the incidence of aberrant cells by comparison with the 1986 figures. Moreover, the incidence of chromosomal aberrations was found to be 50% greater among the mothers than among the infants.

Also discovered was an increase in the radiation-sensitivity of children's lymphocytes in Gomel province. A dose of 0.3–1.0 Gy of gamma radiation proved to have the same effect in in vitro irradiation at the post-synaptic stage of the cell cycle of children's lymphocytes from Gomel province as on children living in ecologically unaffected areas.[29] The incidence of development defects in the southern regions of Gomel and Mogilev provinces $(10.08 \pm 2.3\%)$ definitely exceeds the control figure ($P < 0.05$, that is less than 5%). A careful study is needed to clarify the causes of the variety of discovered defects.

These changes in function, metabolism, etc., can be the basis for the development of various pathological conditions. An analysis of data supplied by health care institutions shows that the regions in question recorded an increase among the adult population of heart and blood vessel disorders, illness associated with hypertension, ischemic heart disease, myocardic heart attacks, respiratory disorders, chronic nonspecific [30] pneumonia.

Children show an increased incidence of chronic inflammatory illnesses of the respiratory tract, of neurocirculatory dystonia, anemia, thyroid disorders, and pathologies of lymphatic tissues. Neurotic and similar conditions are tending to become more common.

Figures show an increase in primary invalidity among manual and office workers, the increase among agricultural workers being smaller. An analysis of the separate nosological groups shows a growth of invalidity in 1986–87 in the contaminated regions resulting from ischemic heart disease, strokes, chronic respiratory disorders, and a drop in the incidence of hypertension (though the number of strokes increased).

Without doubt, one reason for the worsening health statistics is that an improvement in medical services has brought more illnesses to light. However, we must consider firstly the objective deviations in the structure and function of a number of systems (immune, blood, endocrine, etc.) that have been revealed by laboratory methods. Secondly, we must consider the proven steady annual worsening of the health figures over the last 3 years. Thirdly, there is a uniformity in the changes seen in animals in the contaminated regions and in the people who have been studied. All these factors make it impossible to

---

[29] The implications of these data seem to be that children living in an ecologically "clean" environment outside the Gomel region, but eating "dirty" food, receive about the same dose as those living in a "dirty" environment, but eating at least partially clean food.

[30] The term "nonspecific" is in this context used to indicate that the physicians find it hard to understand the observed forms of pneumonia, say, in terms of the traditional diagnostics and causes.

exclude a direct relationship between the radiological situation and the health of the population. To confirm this we can point to the data resulting from the epidemiological research conducted by the different institutions of the Byelorussian Soviet Republic.[31] This study began in the Narovlya region in 1984. It shows an increase in the incidence of ischemic heart disease among agricultural machine operators in the 40–50–59 age groups from 8.8% in 1984–85 to 20.3%. These groups also showed a fairly marked increase in the incidence of arterial hypertonia (from 24.4% to 40.1%).

The acceptance of the proposed guideline for a "safe" lifetime radiation dose of 35 rem gives cause for concern. Until now, such a dose has never been accepted in cases of constant external and internal irradiation – and even the most fundamental aspects of the effect of small doses on human and animal organisms have yet to be clarified. Nothing is known about the remote consequences of their possible treatment. The data obtained by the Institute of Radiobiology of the Byelorussian Academy of Sciences concerning the effect of the irradiated ecological environment indicate that in this situation doses produce the same biological effects as would be produced by much higher levels of purely external irradiation of the organism.

The situation is made worse by the fact that, while accepting this dose as permissible, the health authorities have done nothing to ensure that it is not exceeded. They have, most importantly, failed to regulate working conditions and to check the doses people continue to receive. This last is quite a complicated matter. Along with the well-known problems it is necessary to take into account the presence of isotopes giving off alpha and beta radiation – for which no permitted dose has been worked out. Checks on these types of radiation are needed, especially of their penetration by inhalation, as well as an assessment of their collective effect on the organism. Of crucial importance is the effect of the accompanying ecological factors which could increase the harmful effect of ionizing radiation.

The calculations carried out so far relating to dose levels indicate that the established permissible dose could be exceeded. It is appropriate to note that a cesium-137 contamination level of $10 - 15$ Ci/km^2 can be regarded as the maximum possible for areas in which an unrestricted, normal lifestyle would not lead to a total dose of more than 35 rem.

3. Issues Related to Plant and Animal Husbandry

As has already been indicated, it is important for plant husbandry and, consequently, for animal husbandry that not only the density of contamination of an area should be established, but also that of the individual farms in which the radionuclides are found, and that the mobility of the nuclides in various

---

[31] ByelNIIET, i.e., a Byelorussian Medical Research Institute, and OTI MinObEs, a department of the Ministry of Social Welfare.

types of soil should be understood. According to the data collected by the Byelorussian State Agricultural Institute the percentage forms of radionuclides that can easily (chemically) be exchanged against others (stable ones) is higher on arable land.

It is also important to establish the mechanisms by which the products of plant husbandry are contaminated – both via the roots and via other parts of the plant. Incidentally, while access via other parts was the norm in the first year after the accident, in the following year most of the contamination entered via the roots. In 1987, the non-root portion of total cesium-137 contamination in corn (maize) was 38%, 20% in potatoes, 27% in perennial herbs, 15% in vetch and oats. Non-root figures for strontium-90 contamination are: 22% in wheat, 30% in barley, 10% in corn (25-day seedlings).

A third important indicator is the coefficient of accumulation. Its value for strontium-90 and cesium-137 can vary by an order of magnitude from region to region. The higher figures have been obtained in plants growing on soil which has been ploughed, while the lower figures are for plants growing on unploughed soil, especially in the 30-km Zone. This is explained by the fact that the top layer of soil was not disturbed and, therefore, the contact of the roots with radioactive substances was reduced.

A large source of contaminated fodder is natural hayfields and pastures. Here the level of contamination in the plant cover depends to a greater extent on soil type, on moisture level, and on the land's state of cultivation. For example, it is impossible to obtain "clean" milk on peaty, swampy, podzolized clay soils even when the contamination is only 1 Ci/km^2. "Clean" meat cannot be produced on such soils when the level is 4 Ci/km^2. On soils with a peat bog base the corresponding figure is 4 to 12 Ci/km^2. However, when the very same type of soil underlies acidic pasture land it is possible to produce clean milk even though contamination density is as high as 20 Ci/km^2, and clean meat up to a level of 30 Ci/km^2.

In general, however, taking into account strontium-90 contamination, it is not possible to produce clean fodder or animal products from natural pasture where the cesium-137/134 contamination exceeds 15 to 20 Ci/km^2.

We should note that all the regulatory documents consider only the presence of cesium-137/134. There exist maps drawn up by a Byelorussian research institute (ByelNIIPA) and the Byelorussian Agricultural Chemistry Institute which show levels of strontium-90 contamination. According to these maps, the area of land where contamination with strontium-90 exceeds 0.3 Ci/km^2 in the province of Mogilev is 77 000 hectares and, in Gomel province, 386 000 hectares of agricultural land.

The contamination of plant foods leads to the secondary contamination of animal products: milk and meat. The average trend in Gomel (second column) and in Mogilev (third column) province for milk shows decreasing contamination as the years pass by:

1986 – 86.3% – 46.2 %
1987 – 29.6% – 08.0 %
1988 – 17.0% – 07.0 % .

However, the percentage of "dirty" milk (as opposed to "clean" milk) from the regions contaminated with radionuclides is high. For four regions (Kortyukovo, Cherikovo, Slavgorod, and Krasnopol) in Mogilev province, it fluctuates from 14% to 50% , while in the Bragino, Vetkovo, Narovlya, and Khoiniki regions of Gomel province, it varies from 60% to 66%. For meat produced in Gomel and Mogilev provinces, the figures are: 1986 – 17 500 tons; 1987 – 6 900 tons; 1988 – 1 500 tons. Along with this, animal fodder containing high concentrations of radionuclides and the keeping of animals on contaminated land have a definite effect on the organism's functional systems – the endocrine, immune, and blood cell replacement systems, among others.

From 1986 onwards, according to data collected by the State Agricultural institutions of Byelorussia, an increased number of cows on farms in the contaminated zone suffered abnormalities in the reproductive cycle in the course of pregnancy and calving. A rise was also noted at that time in the number of cows failing to conceive after several inseminations and remaining barren for long periods because of lack of ovulation, because of the persistence of the corpus luteum (yellow body), because of hyperfunction of the ovaries, and other gynocological disorders. There has also been an increase in the number of abortions and of placenta detachment.

Among animals with damaged thyroid glands milk yield dropped to a half or third of that of a control group ("Oktyabr" Soviet Farm) in the first year to 18 months after the accident.

4.   Conclusions

1) The radiation situation in the republic remains difficult and will not change substantially in the near future. It is essential to carry on with the work of building up a clear picture of the radiation problem and its effects on various plant and animal organisms, as well as on humans. It is also necessary to facilitate and expand the work being done, not only on cesium-137 and 134, but also on the isotopes of strontium and other transuranic elements.

2) There are several reasons why extra measures need to be taken. They include:
   – the long-term persistence of radiation in the environment which will lead to the contamination of various ecosystems and of agricultural products, to functional and structural damage in plants, animals, and humans;

– the ineffectiveness (as shown by studies over 3 years) of the approach
adopted in rectification work after the Chernobyl accident;
– the large scale of the accident's effects on the Byelorussian republic;
– the need for sustained, long-term assistance to the affected regions;
– the lack of clarity as to long-term consequences.

Primary among the measures that are needed at this stage is a transfer of population from territories where the cesium-137 contamination is 15 Ci/km^2 or higher. In areas where the level is lower the simultaneous presence of all the isotopes must be taken into account.

3) We should focus attention on living conditions, agriculture, and social welfare in the areas where cesium-137 contamination reaches 15 Ci/km^2. These areas produce contaminated food which is eaten without restrictions by the local people who, in this and other ways, receive appreciable doses of radiation. We also need to mark off those areas where cesium-137 contamination is in the $5 - 6 - 7 \text{ Ci/km}^2$ range and clarify, as soon as possible, what other radionuclides are also present. We need to lay down safety guidelines for people working and living in such areas, while devising agricultural techniques which will prevent the contamination of food products.

4) The experimental and clinical data that we have, concerning damage to the most vital systems of the organism, should warn us of the danger to the health of a section of the population and to future generations. The data show a tendency for the health of those living in contaminated areas to decline and, although there are various views about this, there is no reliable basis for any forecast of future health trends, especially where chronic irradiation amounting to a dose of less than 35 rem is concerned. We must remember the increased number of cases of leukemia, cancer, and other incurable illnesses. This is a well-known fact, but seems not to worry anyone in authority. Many ordinary people, on the other hand, suffer from tense psycho-emotional stress. No attempt has been made to create safe working conditions or to keep a check on the build-up of radiation doses. The health authorities need to adopt a much more energetic approach to the situation which is taking shape.

1.7 Narodichi [32]

One of the areas of the Ukraine most strongly affected by the accident is Zhitomir province. Because no immediate information about the danger was provided, because no prophylactic iodine tablets were distributed, the children of Narodichi received a radiation dose to their thyroid gland of 30 to 2500 rad. The incidence of illness is increasing among inhabitants of the contaminated areas and consumers of "dirty" foodstuffs. The most commonly reported afflictions are cardiovascular, lymphoid, and oncological. The number of children suffering from cataracts has risen dramatically.

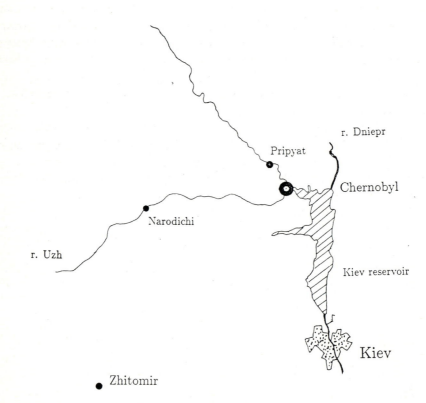

Fig. 1.9. The location of Narodichi region is indicated

---

[32] Chapters 10 and 11 elaborate on the problems referred to in this Section.

*Bogdan Antonovich Korzhanovskii, surgeon at the regional hospital in
the settlement of Narodichi, Zhitomir province, Ukraine:*[33]

I am the head of the surgery unit. I would like to raise the question of children's health. We can divide the 5000 children who live in our settlement into five groups according to the *level of irradiation of the thyroid with radioactive iodine:* [34]

group A (0 –   30 rad): 1473 children;
group B (30 –   75 rad): 1177 children;
group C (75 –   200 rad): 826 children;
group D (200 –   500 rad): 574 children;
group E (500 – 2500 rad): 467 children.

These are the doses the children received in the first days of the accident... just the dose that reached the little thyroid gland. What about a small child's whole body? How will that respond to radiation damage? What sort of dose did the whole body get? And it wasn't just iodine. There were heavy metals, too, which we weren't told about for nearly a year. Then we got these figures nearly 2 years later – and all the while our children were untreated. Now, almost every other child has hyperplasia of the thyroid gland. And all the children, whether they're in the clean or the "dirty" (contaminated) zone have all got lymphodenitis, that is, swelling of the lymphatic glands.

While a diagnosis of *pathology of lymphatic tissue* was made on the cases of four young patients in 1985, in 1987 there were 221. In adults, we have seen an increase of *ischemic heart disease:* we had 518 sufferers by the end of 1986, 757 in 1987, 807 in 1988. In just the first quarter of 1989, we have had 430 of them.

Here are the figures for *cataracts of children:*

1984 –   24 cases;
1985 –   65 cases;
1986 – 178 cases;
1987 – 185 cases;
1988 –   59 cases (for part of the year);[35]
1989 – 195 cases (within the first 3 months).

Now I want to say, as a surgeon working full-time in a hospital – people are suffering changes in their immune systems. People now cope very poorly even with the slightest infections. They recover very slowly from pneumonia. Oncological conditions are on the increase. In my department at the moment there are eight people with cancer of the mouth. These people are in terrible

---

[33] The author conducted this interview in Narodichi on May 16, 1989.

[34] The unit of the energy dose (Rad: radiation absorbed dose) is measured in [rd] or [rad]. Since 1985 the unit Gray [Gy] is used. 1 rd = 0.01 Gy; see also Appendix A.

[35] The eye specialist was ill for a long time, during this time the children were not examined.

condition. We can't even look them straight in the eye. It's terrible. We didn't have cases like that before. And these are young people – under 50. We, as doctors, find this horrifying.

We see a lot of children who are suffering with things we find it difficult to diagnose. Just the other day, we had some children with abdominal pains. What causes these pains, what sort of pains are they? The test results are normal but the children are in pain – shouting, groaning, shrieking. They groan just like adults. These are little children of 3 and 4. We are mystified.

We have no immunology lab. We have a lot of trouble with septic infections because the body just can't fight them.

The incidence of cancer of the mouth is going up among the people who live in the contaminated areas. This is because they eat what they grow and they shouldn't – it's all "dirty". We have hardly any clean milk around here – and they don't bring much in. The shops don't get much clean food. There's no fruit at all, very little fish or juice.

That's why we've seen such a rise in the number of diseases of the mouth, esophagus, and stomach. We must do something. We've got to save these people.

Fig. 1.10. Children from the Chernobyl region shown leaving for a Pioneer (youth organization) camp (Photo taken May 1986.)

1.8 Criminal Actions [36]

After 4 years of silence about the scale and consequences of the disaster, and with absolute censorship of any press mentioning radiation, towards the end of 1989 some timid strains of truth managed to leak through.

The people have demanded that the pivotal responsibility for the accident and the poor handling of it be placed on high-ranking officials, and that their actions should be declared criminal, since they resulted in great suffering for the citizens of the affected areas. These voices of protest against a policy that, in effect, has resulted in genocide are joined by an evergrowing number of USSR People's Deputies newly elected to the various government organs.

Vasilii Yakovenko, a writer, member of the Central Committee of the Byelorussian Communist Party and of the Council of Ministers of the Byelorussian Republic, member of the presidium of the Byelorussian Ecological Union, Chairman of the Journalists' Section of the Byelorussian Writers' Union, wrote October 27, 1989, in Minsk, Byelorussia, an open letter to the General Procurator (Attorney) of the USSR, A. Ya. Sukhorev, and to the Procurator of the Byelorussian SSR G. S. Tarnavskii. Yakovenko requests that criminal proceedings be instituted against certain persons whose activities, since the beginning of the Chernobyl emergency operation, have been and continue to appear to him to be criminal in character and are likely to have serious consequences for the nation. The well-known Byelorussian journalist's address to the Procurator was supported by a large group of USSR People's Deputies, including the writers A. Adamovich, Ch. Aitmatov, V. Belov, V. Bykov, I. Drutse, Ya. Peters, V. Rasputin, Yu. Chernichenko, Yu. Shcherbak, V. Yavorivskii. Signed October 27, 1989, Minsk, Byelorussia.[37]

The Chernobyl accident has brought unparalleled suffering not experienced (since World War II) to the Byelorussian people, as it has to the inhabitants of a number of provinces of Russia and the Ukraine. The very real danger made it necessary to evacuate the population from the 30-km Zone. The result was a massive movement of people suffering from radiation sickness. We saw

---

[36] Chapters 9 – 12 further detail the problems cited in this section.

[37] V. Yakovenko provided the author with a copy of this open letter at the beginning of 1990.

refugees again – it was a moment comparable in its tragedy to wartime. But the "front" was not to be seen, only felt as a metallic taste in the mouth.

From the first days of the tragedy we clearly saw constant attempts to conceal this insidious, universal woe from our own people. We saw a totally indifferent attitude towards the people's health.

Fearing to upset the people with an uncompromising, explosive revelation of the truth at the time of the accident, the Byelorussian Ministry of Health and the Minsk City Executive Committee, through its leaders and those of the medical profession, did nothing to protect the people of the capital from the radioactive cloud.

Secrecy at a time of desperate adversity only multiplies the suffering involved.

In March 1987 a group of experts from the USSR State Agricultural Institute and from the Agricultural Institute and Ministry of Health of the Byelorussian Republic laid a proposal before the USSR Council of Ministers. They urged that agricultural activity cease and that the population be evacuated from the zone where contamination exceeds 40 Ci/km^2. This proposal was, however, rejected by the chairman of the Chernobyl emergency commission B.Ye. Shcherbina. He even declared secret the details of contamination in individual towns and villages, where it was above the permissible maximum.

This declaration that the facts were "top secret" permitted certain high-ranking officials the luxury of not knowing. "Not to know" is a very convenient mode of existence. But ignorance, just like inaction, becomes criminal when indulged in by individuals who enjoy the people's trust.

According to the evidence given by doctors throughout the affected regions, there has been, especially among the children, an increase in the incidence of chronic diseases of the nose and throat, of the stomach and intestines, of the liver, of the spleen, and other organs – even of the blood. According to the official statistics, which the local specialists believe to understate the problem, the state of health of half the children living in the provinces of Gomel and Mogilev is unsatisfactory. Their blood lacks essential elements. In other words, it is thin and lifeless, causing the body's defenses to be weakened and its natural immunity to be undermined, as it is in cases of AIDS. Perhaps that is what it is – radiation AIDS? The doctors, however, are still playing politics and are unwilling to connect the children's condition with the Chernobyl accident. Probably, complying with the instructions issued by the medical authorities, they do not even allow such thoughts into their heads. They should. The connection is obvious, whether it is direct or indirect. Restricted in their outdoor activities, sometimes locked up indoors (whether at home or school), the children are also not fed properly because local food is "dirty" and deliveries of clean food, especially vegetables and fruit, are insufficient. Hence, their condition, which is close to anemia, if not anemia itself. Laboratory studies have shown that the people have developed a susceptibility to "auto-immune pathology".

Adults in these areas complain mainly of headaches and dizziness. According to figures relating to the Narovlya region which have been compiled by medical experts studying the effect of radiation on work-fitness, in 20 cases out of 100 a "low level of physical activity" was recorded, something never observed there before the Chernobyl accident. The researchers also found a significantly greater incidence of heart and blood vessel disorders among agricultural machine operators. They noted the susceptibility of the population as a whole to strokes, acute conditions of the ear, nose, and throat organs, and arterial hypertension.

Here are the *cancer statistics for the Slavgorod region*, with the number of cases following the year:

1981 – 14;
1982 – 12;
1985 – 11;
1986 – 25;
1987 – 48;

1988 – 70;
1989 – 34 (first 6 months only).

These figures are made even more horrific by the fact that over the last 2 or 3 years the population of the region has gone down by about 20%. The same trend can be seen in the cancer figures for other regions, including Cherikovo, Krasnopol and Vetka.

The deformed animals that have been born have been shown in quite a number of films and television programs. During the last few weeks, I have visited many of the "dirty" areas and I know just how serious a matter this is. Here are the figures for *deformed animals in the Slavgorod region* of Mogilev province:

1985 –   5;
1986 – 21;
1987 – 39;
1988 – 84;
1989 – 50 (first 7 months only).

The picture is roughly the same in the regions of Khoiniki and Vetkovo. Calves are born without teeth, without extremities, without hair, with incomplete digestive systems. Hematological studies of the blood of cattle have revealed a predisposition to leukosis.

All over the contaminated provinces of Byelorussia there is evidence of a decline in blood vitality, of genetic damage, of increased numbers of abnormalities. All this can only lead to mental and physical debilitation, to an epidemic of "radiation AIDS", to the degeneration of the whole nation. It is made worse by the fact that radioactive dust is spread by various means, knowingly and unknowingly, throughout the republic. Is it possible that, without the connivance and encouragement of the medical services, things like the following could happen?

- Millions of roubles have been invested in decontamination stations along the roads of Mogilev province, but construction work has stopped and plans to bring them into operation have been suspended.
- Machines (tractors, cars, trucks, combine harvesters, etc.) taken out of use because of heavy contamination and due to be scrapped, have been taken from Bragino, Khoiniki and other contaminated regions to various parts of other regions which are comparatively or completely clean.
- The refrigeration rooms of meat-processing plants are still full to this day of "dirty" carcasses from 1986, and tens of thousands of tons of this dangerous meat have been mixed in with clean meat in food products.
- Contaminated animal fodder, including mixed feedcake and hay, has been taken from the evacuated zone and distributed around the neighboring villages and regions of Gomel province.
- In defiance of the guidelines laid down by the Byelorussian State Agricultural Institute, the Gomel Agricultural Committee ordered (order No.138, 23rd April 1987) that land in the Bragino, Khoiniki, and Narovlya regions, which was condemned as unfit for use, should be returned to cultivation. This land has been sown by teams from practically all the regions of the province and fabulous bonuses have been earned by the people who have produced harvests from it – sometimes amounting to six or eight times their normal monthly salary.
- Regional processing plants have brought abandoned materials out from behind the barbed wire of the restricted zone and have used them to double their planned output. They have shared their profits with certain officials.
- Dirty cattle are taken to clean areas for fattening. Through their manure they constantly bring contamination to new places.
- The skimmed residue remaining after the processing of "dirty" milk is returned to the farms and fed to calves.
- Within the republic and also beyond its boundaries mass production, processing and delivery of various products such as turf briquettes, firewood, and timber continues.
- School children have been taken to forest plantations in the zone.
- The *Directive on Agricultural Activity in Conditions of Radioactive Contamination* issued by the USSR State Agricultural Institute in 1988 has not been adhered to

1) as regards the ban on agricultural production where contamination density exceeds 80 Ci/km^2 (production continued in Krasnopol and other regions as late as in 1989);

2) as regards the removal from the rotation cycle of arable land with a radiation level exceeding 40 Ci/km^2: according to the radiologists on the spot the Directive was only implemented in Vetkovo a year after it was issued.

A new phase of the "Chernobyl Cleanup" began in the spring of 1989 with urgent demands from the public, from journalists, writers, and scientists that the Chernobyl hypocrites should be unmasked and that a more radical approach to the protection of the population from radiation should be adopted.

I consider that the people and its legal institutions have reached the point where they are ready to call to account those involved in the Chernobyl emergency operation and to charge them with

a) sustained and premeditated concealment of the true situation in the contaminated areas;

b) such indifference and negligence in the defense of what is most precious – our health – that damage has been inflicted on our people;

c) the faking of pseudo-scientific "guidelines" allowing unrestricted residence in contaminated territory;

d) the abuse of power;

e) the waste of hundreds of millions of roubles in State resources on construction projects which were absurd from the moment of conception.

The people and the nation cannot prosper unless our judicial institutions take action!

Fig. 1.11. The sign reads "AFFECTED" (by radiation)

1.9 "Children of Chernobyl"[38]

After the hot summer of 1986, spent working in the Special Zone, it was necessary for me to spend some time in the most contaminated parts of Byelorussia, in Gomel and Mogilevsk provinces, and in Khoiniki, Braginsk, and Narovlya regions. I was appalled by the condition of the children.

There are practically no healthy children in any of the towns or settlements there. The appearance of ambulances on school grounds is a common sight. Children faint from weakness right in the classrooms.

The women have organized themselves into committees called "Children of Chernobyl" to help sick children. In response to the terrific rise in the number of sick children, the mothers have pleaded for aid from the highest levels of government.

In April, 1990, from:
The Byelorussian
Women's Action Committee
Gomel
Byelorussia [39]

To Comrades
M. S. Gorbachev,
N. I. Ryzhkov,
and to all Deputies of the Supreme Soviet

"Dear Mikhail Sergeyevich, Nikolai Ivanovich, and Deputies:

We address you in the name of the women of Byelorussia. We are entering the fourth year of the greatest tragedy of our age – the tragedy of Chernobyl. Nevertheless, there has still been no evacuation of the people, not even of the children, living in the contaminated villages of the Bragino, Narovlya, Buda-Koshelevo, and Dobruzh regions of the province of Gomel; nor from the Kostyukovichi, Cherikovo, Krasnopol, Slavgorod, Klichi, Bykovo, and Klimovichi regions of Mogilev province; nor from the Volozhino region of Minsk province; nor from several villages of the province of Brest where the cesium-137 contamination level is 30 to 146 Ci/km^2 and the strontium-90 level is 1 to 1.5 Ci/km^2.

---

[38] See also Chapter 11.

[39] A copy of this letter was given to the author by representatives of the Byelorussian branch of the union "Children of Chernobyl" in Mogilev in 1990.

Of the three republics affected by the accident, Byelorussia has borne the brunt of the tragedy.

Even according to official documents, approximately 70% of the radioactive cesium ejected from the ruined reactor fell on Byelorussian territory. As a result, practically all the citizens of our republic live constantly in a radioactive environment.

Any dose of radiation is dangerous – large or small. Its effect is seen in the weakening of resistance to various illnesses, in the breakdown of vital physiological systems, in the weakening of the immune system (which prevents development of tumors) and in serious genetic damage. All these have appeared in Byelorussia. For example, in the Khoiniki region, out of the 200 children that have been born, 30 were born deformed.

Official figures for cancer cases in the provinces in question show in 1986–88 an average increase of 14–20% compared with 1981–85.

In the regions listed above there are more than 63 000 children, of whom 36 000 are under school age. Since the Chernobyl accident these children have suffered on a massive scale from anemia, reduced immunity, tonsilitis, bronchitis and unusually prolonged pneumonia. There is no future for the children or the adults who live in the high-radioactivity areas!

It is horrible to see a child, its face more like a white mask with black bags under its eyes, sitting on ground where the contamination level is 60 Ci/km^2.

Since April 26, 1986, the Chernobyl disaster has been with us every single day, and it will continue until the last person is evacuated from the contaminated lands and trees are planted there instead.

Cries for help from people living there are lost amid the disinformation and the sweet words which seek to avert conflict, amid the empty promises poured out by those who govern our public. The strikes at Narovlya go unreported! They are simply ignored. The cruel truth is made known only through the actions of the People's Front of Byelorussia and not by the functionaries, appointed against the people's will, who wield power.

We believe that these same people, defending their own privileges, tell only half-truths even to you. We want an answer to this question: Why are people still living in the contaminated areas? They not only live there, they produce food which they eat themselves and supply to others. Who is carrying out this vile, deeply horrific experiment on us? Who is planning a future in which the Byelorussians are a nation wiped from the face of the Earth?

The continued food production in the regions of Byelorussia, which are polluted with radionuclides, not only increases the radiation doses suffered by the people who live and work there, but also leads to the distribution of food containing radioactive cesium, strontium, plutonium, and other nuclides in clean areas of Byelorussia and in the other Soviet republics.

As a result, the threat to health spreads wider and wider. People in positions of responsibility in the USSR Ministry of Health, instead of showing concern for the health of the people, accuse us of radiophobia. In April 1990, it will be 4 years since the disaster of the century befell us. Only recently, with inexcusable tardiness and under public pressure, did the Supreme Soviet of the Byelorussian Republic accept an emergency program of rectification.

Do not delay its acceptance! It is no good waiting another 4 years while the Union program is prepared. For the Byelorussian people this will mean the horror of illness, deformity, the extinction of a nation which once contributed a rich culture to the world.

We, the inhabitants of Byelorussia, therefore petition you to take up the following issues, the resolution of which cannot be postponed. We ask you to:

1) Confirm the Byelorussian Supreme Soviet's emergency rectification program and order the immediate evacuation of the population from the contaminated areas.

2) Until the program is carried out, to transfer mothers with children to sanatoria run by Section No.4 in Byelorussia and the other republics.

3) To ban farming in the contaminated areas and, after 1990, to stop contributions of food from these areas to national stocks.

4) To plant forests in the evacuated areas.

5) To stop the practice of supplying the leadership of the KGB, the Council of Ministers, the State Planning Committee, the Agricultural Institute, the Ministry of Health, as well as the officials of provincial and regional governments with higher-quality food via the network of special farms, shops, cafes, and restaurants. The leadership of the republic must show by its deeds that it truly represents the people. This must be enforced by law and by the vigilance of the people. We believe that this measure will make it easier to keep a check on food supplies in the republic and at the same time an infringement of social justice will be removed.

6) The special meat processing plants, bakeries, and confectionery factories which produce foods for the higher echelons of the administration must be turned over to the production of clean food for children. The milk and meat produced by the special farms should be dealt with in the same way. Food should be issued to children according to medical certificate. Since these measures will not solve completely the problems involved in feeding the children, more food must be bought abroad.

7) Allocate new premises for children's hospitals in Minsk, Gomel, and Mogilev, and supply them with the necessary modern equipment. Organize diagnostic centers in the provinces of Gomel and Mogilev.

8) Supply the republic's population with proper radiation counters for individual use in the examination of food products delivered from the fields and farms. Supply the clinics, especially those in the contaminated areas, with disposable syringes.

9) The negligence and incompetence of some should not be concealed by the patriotism of others. Therefore, the despatch of young people, experts, students, and military personnel to contaminated areas to carry out agricultural and decontamination work should be banned. This practice can only lead to the degeneration of the whole nation.

10) To remove from sale all food stuffs which fail to pass radiation tests or which pass only by the criteria laid down for wartime.

11) Make available, as soon as possible, resources from central funds to tackle the problems listed above. Since the Atomic Energy Authority was responsible for the accident, let it now suspend the construction of its 11 nuclear power stations. The money thus saved should go to help their victims, the people of the contaminated regions of Byelorussia, the Ukraine, and the Smolensk province of the RSFSR.

We make this plea not only out of concern for our own health, but out of concern for our future and the future of our children.

Comrade Deputies, to fail to understand this, to fail to take the action which can no longer be postponed, would be a crime which our descendants should never forgive."

1.10 The Rectification Program's Summary of the Situation [40]

The dimensions of the contaminated territories, the cost and, what is most frightening, the number of people who remain on territories where a healthy life is impossible, are staggering. Towards the end of the fifth year after the catastrophe, finally, federal and republic programs were developed to combat the consequences of the accident.

These programs are foot-dragging, poorly financed, and have little scientific basis. Even official information sources state that, in Byelorussia alone, 353 settlements are waiting to be evacuated; in the Ukraine there are 67 such places.

Fig. 1.12. The location of the radioactively polluted region in relation to Moscow

---

[40] For further details see Chapter 15.

From the Chernobyl Rectification Program for 1990–1995 (Moscow 1990), based on the results of a working group founded by the Government Commission in Chernobyl on March 2, 1990.

The accident which took place at the Chernobyl nuclear power station on April 26, 1986, was, in its scale and in the damage it inflicted, one of the most serious catastrophes that have ever happened on our planet. It resulted in the ejection of a considerable mass of radioactive substances into the environment, with grievous ecological consequences. The situation on May 10, 1986, was as follows:

$\simeq$ 1100 km^2 of territory where the radiation level reached 20 mR/h;

$\simeq$ 3000 km^2 where the level was 5 mR/h;

$\simeq$ 8000 km^2 with 2 mR/h.

At the beginning of 1990 the situation was as follows:

$\simeq$ 3200 km^2 where contamination density reaches or exceeded 40 Ci/km^2;

$\simeq$ 7500 km^2 where contamination density is $15 - 40$ Ci/km^2;

$\simeq$ 14 000 km^2 where contamination density is $5 - 15$ Ci/km^2;

$\simeq$ 76 100 km^2 where contamination density was $1 - 5$ Ci/km^2.

$\simeq$ 800 000 people still living in areas where contamination exceeds 5 Ci/km^2;

$\simeq$ 3 070 000 people living in areas where contamination is $1 - 5$ Ci/km^2.

The accident disrupted normal life and work in a number of regions of the Ukraine, Byelorussian and the Russian Federation.

144 000 hectares of agricultural land were taken out of cultivation. The work of many factories and farms was halted. In monetary terms initial losses amounted to 900 million roubles. Loss of agricultural production and losses in other industries where production was interrupted is estimated at 1200 million roubles for the period 1986–89. Payments from the State Insurance Fund to citizens, farms, and cooperatives totalled 274 million roubles.

116 000 people were evacuated from the danger zone.[41] The total sum spent on emergency rectification work in 1986–89 was, together with the cost of the damage done, 9.2 billion roubles.

The sum spent by the Ministry of Defense on emergency work following the accident – mainly on decontamination of installations and settlements – totalled, in 1986–89, 577 million roubles.

The Council of Ministers of the Republics have presented emergency aid programs for rectification work in the RSFSR, the Ukraine, and Byelorussia during 1990–95 to the State Council of Ministers. The total planned expenditure is 33 billion roubles.

These programs include details of measures to be taken and work to be completed in the period to 1993. Expenditure during this period will be 16 billion roubles.

---

[41] The 30-km Zone.

Taking into account the specific details of the republics' programs and the need to avoid duplication of efforts, a program of rectification works to be carried out in the Chernobyl restricted zone during 1991–95 has been prepared at the government's request by the Ministry of Atomic Energy. These measures will cost in the region of 2.4 billion roubles.

The programs prepared by the Byelorussian and Ukrainian republics envisage the gradual transfer of all inhabitants now living in areas where the level of cesium-137 contamination exceeds $10 - 15$ Ci/km^2. In Byelorussia this stage of the transfer will affect 96 731 people living in 353 towns and villages. In the Ukraine 29 731 people will be moved from 67 settlements.

Fig. 1.13. Rectification work. (Photo taken June 5, 1986.)

1.11 The Chernobyl Union [42]

Then, there is the tragic fate of the persons who worked in the
30-km zone of the Chernobyl power station between 1986 and
1990, the liquidators (or rectifiers) as they are called by the press.
There are between 650 000 and 1 000 000 such persons. They
worked in the areas of highest contamination – without appropriate
medical monitoring or radiation protection. These are the people
who collected the radioactive graphite using nothing but shovels,
who built the Sarcophagus, and who hung over the hulk of the
burned-out reactor dome in helicopters. In 1986, these people were
cheered by the whole nation, they were called heroes.

Now they are of no use to anyone. The government has
shrugged them off. Their illnesses have been declared "not related to radiation", not connected to their working in the deadly
radiation zones.

> *Georgii Lepin, 53 years old, a professor (Doctor of Science). He is*
> *one of the founders and vice-president of the all-union voluntary or-*
> *ganization "Chernobyl Union" (registered February 14, 1990 with*
> *the Ukrainian Council of Ministers) and has been in Chernobyl since*
> *November 1986.*

Now you often hear cries of "Chernobyl! Chernobyl!". Just how much can you
say and write about it? I think that Chernobyl is still waiting for its Solzhenitsyn,
who will tell the truth about it.

The worst thing is that Solzhenitsyn described the past. The relatively dis-
tant past, but this is now. Today.

Take, for example, that silent lie which cost the people of Pripyat and the
villages around it so dearly. The only explanation for what went on in Pripyat
during the first 36 hours after the accident is that it was an attempt to conceal
what had happened, an attempt to avoid glasnost. Those who finally took the
decision to evacuate the inhabitants should not imagine that they saved them. It
was luck that saved them – the fact that the "tongue" of debris from the reactor
passed just south of the town, through the "Red Wood". It was this forest area
that bore the first, the most savage blow. If the wind had changed direction
just slightly then those 36 hours would have been the last hours in the lives of
a multitude of people.

---

[42] See also Chapters 4 – 7.

Even as it was, the fact that the game of "I know nothing, I have heard nothing and I'll say nothing" was played for so long was quite bad enough, as far as its consequences for the people of Pripyat were concerned.

It is true that the Pripyat people often say that the leaders at the province and republic level did not forget about them, they were just unable to do two things at once. First, there was an urgent need to evacuate some other people from a much more dangerous place – that is, their relatives had to be got out of Kiev. Then it was Pripyat's turn.

They say that there are such things as "lies for the sake of a greater good", "lies that save", even "sacred lies". Maybe there are. But was this such a case? Can it be that we are so weak that we cannot face the truth about ourselves, even if it is often very bitter? When shall we finally understand that further progress is unthinkable without the strength that only the truth can give?

There has been no consideration of the fact that people who have taken part in the emergency rectification work will suffer serious consequences. This has hardly been mentioned.

It is sufficient just to examine our newspapers. Where will you find in print the number of people who died after Chernobyl? You might find a figure of 30 people. That's all. The 31st won't be there. But I could give a different figure. Between 5000 and 7000 Chernobyl people have died so far. They were scattered all over the USSR. We can't count them accurately. But on the basis of the evidence supplied periodically by our organization the figure can be estimated.[43]

And what about that other big lie, which seems to have the backing of the Ministry of Health itself, the lie that comes with the stamp: "Not radiation-related". How many people are there, seriously ill as a result of the accident, who cannot get this dirty, inhuman stamp removed?

When I tried to ask at our local surgery what the people who were at Chernobyl are dying of, the response was: "What do you mean? It's nothing to do with Chernobyl. They are dying from other causes. Some have heart attacks. Some committed suicide."

But why? We brought people to this, after all. What if a lad of 25 or 30 becomes a complete invalid on a beggarly pension, 70–100 roubles a month? And if he's got a family? What is he to do? There's only one thing – to hang or drown himself.

We estimate that today there are at least 50 000 invalids who took part in the emergency work. Our people in Moscow tried to work it out – they thought

---

[43] Professor Lepin speaks here about the number of deaths out of the about 25 000 who worked in the high-radiation fields. This concerns an age group of up to 35 years. For further data and discussions see also Chapters 5 – 7.

the number was nearer to 100 000. I think the figure of 50 000 is closer to the truth at the moment. This is about 10% of the total number of people who passed through Chernobyl during the emergency operations.

Alas, Mikhail Sergeyevich [Gorbachev] has not honored us humble Chernobyl workers with his attention. We would have a lot to say to him. And it would certainly do some good. As for this "theatrical presentation" that we have been shown on television, it could take its rightful place among the classic works of that period in our history that we are so desperately trying to leave behind.[44]

It is clear that this lie, which takes various forms but is the same one in essence, serves somebody's purpose. It could be that for these "somebodies", the lie is the only possible means of survival.

How paradoxical that honesty, openness, glasnost have become a sort of facade behind which we continue to deal with things according to our old and, oh! how deeply rooted habit, via a "back entrance" which is littered with the rubbish of all kinds of distortions, hypocrisy and show, with outright lies and half-truths.

But listen to the words of the "rectifiers" as they have started to call those who passed through the disaster zone. Try to understand them. Their tragedy will shake you. What can you say, how can you find softer words, more like the words we are used to? It's a tragedy, simply a tragedy.

There lived a man. Life was not easy for him, but nevertheless he was glad to be alive. He rejoiced in the sun, the forest, the waters. He rejoiced in little children, he believed they would have a life better than his own. He rejoiced in his own health, he believed that he would have strength enough to bring happiness to his family, to those close to him. And suddenly something happened in some unheard-of place called Chernobyl. They told him his help was needed. He was there just for a little while, maybe only for a couple of minutes he dashed out onto some "dirty" rooftop or into some other "dirty" place. And what a patriot that made him! How the Motherland would thank him for his self-sacrifice and heroism! Of course the Motherland would not forget him. She was already waiting to look after him, her faithful son. She would not let him stumble on the tortuous bends of life. In chorus and individually his immediate bosses and their bosses swear to be faithful (in the name of the Motherland, of course). And to remove his last doubts they give him a written document (no expense spared) with seals, signatures, and numbers of "insurance certificates". Live, they seem to say, enjoy life, everything is still before you.

---

[44] Lepin is referring to the visit by Mikhail and Raisa Gorbachev to Chernobyl on February 23, 1990. It lasted 40 minutes. After Gorbachev had inspected the shield of Block 2, he left for the town of Slavutich from where the television report on his visit was broadcast.

A stabbing pain, something strange pressing somewhere. No need to over-react to such trifles. It will pass. Every time you feel it more – that everything really is passing. The first thing to pass is your faith in your own health, then your belief that your problems, and your health, too, are of any concern to anyone outside your immediate circle. Then you lose faith in those "insurance certificates" and in the people who were so generous with them. Everything that made life worthwhile in your pre-war (pre-Chernobyl) life is past and gone. Even your faith in life and its justice is gone.

A lot of things that were previously unknown have now entered your life. The fact, for example, that it is difficult to get up to your floor without an elevator (it never occurred to you before!). It is hard to be in hospitals (where do you spend more time? At home or there?). How hard it is to make your miserly pension stretch to things which before you took for granted. But the hardest thing is to look at your family, your children, your wife, your parents, who are losing their faith in life and its justice just as you are.

Years pass. Life, divided by Chernobyl into "before" and "after" skids disastrously and slides off track. Driven to despair by indifference and deprived of essential medical help, the "rectifiers" resort to extreme measures.

A hunger strike in the country's main radiological clinic at Pushche-Volodetse, a suburban resort near Kiev. The miners are on hunger strike. Danger and risk – that was their profession. But the risks of Chernobyl in May and June 1986 were the last risks of their young lives. These young lads, like many thousands of other Chernobylites, have been driven into a dead end, a tunnel with no light at the end. Everything that had built up inside them, alas, for many years, burst out at this moment. Chernobyl simply revealed the nature of

the life these young men had been living long before the accident happened. The fire of Chernobyl helped to light up the dark corners of our systems, the places we for so long tried not to notice, about which we so shamefully kept silent. The myth about how proudly the word "Citizen" rings in our country was reduced to ashes in a moment.

"Doctors! Doctors!" By the third day of the hunger strike this call had gone out five times. An ambulance took one lad away. Another was taken for resuscitation. The rest were treated on the spot. No one doubted for a moment that the hunger strike was mortally dangerous for these young men. What mockery of their human dignity, of their fate and their lives could have brought them to such a decision? And what did they gain by it? In this case again, only small flashes of light in their dark tunnel. Only a small part of what a person should be able to expect in a humane and civilized society.

They did not manage to achieve their main aim. The inhuman attitude towards the victims of Chernobyl still rules supreme. Formerly, healthy people become sick, become unfit for work or die after Chernobyl. The government, however, puts all its effort, not into saving them, but into "defending" the State from these people who honestly did their duty as citizens, who risked their health and life itself by answering the call of this very State. It is hard to consider such a policy towards the victims of the Chernobyl tragedy worthy of a self-respecting State.

Fig. 1.14. The aftermath: a kindergarden in Pripyat. The "snow" in the photograph is caused by the film's reaction to high radiation (Photo taken in May 1986.)

2. The Explosion

"It has been calculated that, for the reactors built in the mid-1970s, the most probable type of accident would be one involving melt-down of the core. Its probable consequences would be: deaths – less than one case; acute radiation sickness – less than one case; long-term ill-effects – less than one case; genetic damage – less than one case; financial loss – 100 000 dollars (not counting damage at the station itself)...

"The risk from radiation is about equal to the probability of being struck by a large meteorite able to penetrate the atmosphere and reach the Earth's surface."

From "Atomic Energy.
Mankind and the Environment"
Academician A. P. Aleksandrov (Ed.)
2nd ed. (Moscow 1987) p. 152

"It was not the reactor block that burned.
The fire was the product of decay –
the fallout of long-accumulated lies."
Vladimir Shovkoshitnyi

Unfortunately, the first direct information about the accident at Chernobyl Nuclear Power Station did not reach society and world's public through an official communication of the authorities responsible for the Chernobyl nuclear power station. It was an alarm raised at a Swedish nuclear power station (as a result of routine measurements outside their reactor) which triggered insistent inquiries which eventually led to a first communication

---

Preceding page, a view of the remains of the exploded Block 4. (Photo taken on April 28, 1986.)

from Soviet officials. Their information tried to convey the spirit that "something has happened, but everything is under control". In the first few days and, apparently, up to now, that seems to be the preferred way of referring to the accident and the related problems.

But was there really no danger whatsoever for people living outside the boundaries of the Soviet Union? Was there no need for them to worry about anything concerning their well-being and safety? According to opinions voiced at that time and up to the present, other countries did have reasons for grave concern. The official attempts were to play down the role of the accident, to indicate that there was no real danger. But from the scarce information provided by Soviet authorities and from independent investigations of the radioactive material ejected into the atmosphere and subsequently deposited onto different parts of Europe, the world knew that the situation must be grave.

In particular, in the first few hours and days the attention of the world press focused on the firemen who were the first to step in to battle the burning reactor, and who were consumed by the radioactive flames. But the reactor staff on duty that fateful night were no less courageous.

"Get water into the core!" was the command from Moscow, and the crew again and again forged into the darkness of the torn up reactor and tried to get at least a trickle of the water out of the fire hose into the active zone, the reactor core. The core which no longer existed. Radioactive graphite thrown out by the explosion was strewn about all over the place.

Block 3 was shut down towards the morning of the 26th, Blocks 1 and 2 on the morning of Sunday, April 27.

That fateful night it was not only Block 4 that burned, but also the plant's workers who were scorched by the flames and the intense radiation fields, the heroes that stayed at their posts to the end.

2.1 An Operator Recalls: Explosion – Shock

What exactly happened at Chernobyl on that dreadful night of April 26, 1986? Oleg Genrikh, 30, an operator working in the central hall of Block 4, gives his account in the following.[1] That night Oleg was working with Tolya Kurguz. The time was 1:20 a.m.

C: Oleg, where were you at that moment?

G: We were up at level 35. Do you know the smoking room in Block 3? There was a room just like that in Block 4. We were there together. I had just finished one shift and was carrying straight on with a second one. Within that room they had divided off another little room. They had moved some of the individual protection gear there nearer to us, because it had been in another place. Tolya said to me "Did you write anything in your notes about the crane-drivers?" I told him I had. He said, "Let me copy it then." We had an exam coming up in 3 or 4 days. I said, "You copy it and I'll have a rest for a bit." I needed to, as I was just starting a second shift. It was hot. I took off my jacket. You know how we used to dress. I had a dosimeter hanging on my jacket. My pass was in my pocket. I went into the little room – and right then came the explosion. A mighty crack. Then a hissing. The lights went out immediately. Water started gushing. I was sort of sealed off, though, in the little room. But Tolya was out in the other room and there was a ventilation pipe above him, a great fat thing. He shouted, "Oh, God, I've been scalded! What's put the lights out?"

We were in the dark. I moved a hand, then a foot. They moved alright – so I was still alive. But when I heard my mate shouting, I started shouting, too. I pushed the door open and dashed over to him. But as I did so a powerful jet of steam hit me in the face and chest. I covered my face with my hands and fell to the floor. It was easier to breathe down there. I started shoutings "Tolya, over here! It's better here!" He crawled over and we lay down in one corner of the little room, on the floor. It was impossible to breathe if you stood up. It was like a sauna, there was still this hissing, and gate valves started slamming. I said, "Tolya, what's going on?" I went into a state of shock, I was terrified. He kept his wits about him, he was that kind of person. He had been working there longer than I had, he was a senior operator. He had more experience. He said, "Hang onto me, we're getting out," then, "Looks like we're up to our necks in it." We started to crawl out. I hung onto him, and at the same time I was holding some sweater that I had grabbed in my shocked state. The door of the room was bolted. Tolya opened the door and we set off, not towards the main hall, but towards Block 3. First of all, we went to the elevators. We crawled

---

[1] This interview was conducted by V.M. Chernousenko (C) while visiting O. Genrikh (G) in Moscow in April 1990.

there and looked. The wall had gone! There were no elevators, either. There was no way forward, just a hole. There were no stairs, either. We started to go through the ventilator block to our colleagues in Block 3. There, everything was in a heap, pipes, the doors were jammed. We crawled through a hole. Everything was in darkness. The only light was the light from outside. There was some kind of slab lying in the corridor, we crawled over it. I held on to Tolya, he crawled in front.

C: Why were you crawling?

G: It was like the end of the world, terror, shock. We couldn't get through the ventilator block. The ventilator block was to the left – to the right was the "clean staircase", as it was called. The emergency lighting was on there. That's where I saw my friend's dirty face and his arms all torn to ribbons. The skin hung off his arms like rags. I was amazed that he had managed to open the door with his hands in that state. He shouted, "It hurts! Let's get going." We ran. On the 27th or 24th level I at last realized I was carrying this sweater, and I thought, "Why the heck am I carrying this? It's no use to me," so I threw it away. When we got down to level 9, we met two gas circuit operators – Igor Simonenko and Volodya Semikopov. I asked them, "What's happened?" I didn't understand what was going on. They looked at me, I must have been quite a sight, and they said, "Hey, cool down, don't go wild!" I was in a state of shock. At that moment Dyatlov ran past, the deputy chief engineer of Phase 2. I reported to him, "There's nobody in the central hall. We are operators from the central hall of Block 4." When he saw the state Tolya was in he said, "Quickly, get our friend here to the sick bay." We ran along level zero. The door into the medical room was blocked. There was no way of getting in. We went further up. Semikopov smashed the glass. We all crawled out into the open air. We picked up Tolya and passed him through to our friends outside. We ran along the whole block on the side facing the Spent Fuel Store. As we ran we could see that the central hall was wrecked. We would have done better to run along level 9 and get out through Administration and Staff Center No.2 (Admin-2). But we had run out of the block. Tolya was shouting, "Help me. The pain's terrible!" You could see that he really was in pain. We were compelled to run outside. I only had one thought, my friend had to be helped. We got to the guardpost at Admin-2 and met the regular soldier who was on duty. The station is guarded by these young regulars. This one stood there with his pistol in his hand and shouted "Halt! Stay where you!" I swore at him and said, "Look at this man! Look at his arms! Just raw meat! Get an ambulance!" At this he put his pistol away. He could see that the situation was serious. Tolya cried out, "I'm thirsty." The soldier ran to get some water. He brought it and started to give it to Tolya. Tolya couldn't hold the glass himself. I saw this concrete lorry leaving the site. I whistled and shouted at the blokes in the cab, "Stop! Stop!" They stopped. I explained why I had stopped them. An ambulance zoomed past with its siren wailing. The gates were open. The first place it passed on the station site was Admin-2. I said, "It's all right, mates, you can go, the ambulance has come". I went back to the guardpost.

The ambulance had turned in and had already driven up to Admin-2. They started lifting Tolya into the ambulance. Naturally, I went to get in with him. Simenenko and Semikopov, who had run to the guardpost with us and argued with the soldier on duty, now saw they were no longer needed and went back to the block, to Admin-2. They went to the place where we always reported for duty. The guard shouted, "Show me your pass!" Tolya pointed to his pocket and the guard took out the pass. But I had left my jacket behind. I had taken it off. I told him my name was Genrikh and that my jacket was back in the block with my pass in it. He said, "Right, you're not going through. You are staying here!" I said to him again, "My name is Genrikh. Write it down. My friend here and I were together when this happened!" I knew that I had had a good dose of radiation, but I didn't know it was such a big dose. I knew that when we were running and we saw the ruins of Block 4 we must have got some. But the regular said, "No, I'm not letting you pass through." Meanwhile, Tolya was shouting, "I can't stand the pain. Let's get on with it." Of course, I had to let the ambulance go which took Tolya away.

I went to Admin-2, to the point we report to when starting work. I took off my dirty clothes. I went and had a wash, using the special protective powder. I got dressed in some clean clothes and put on a simple protective mask. I went to the place where we pick up the dosimeters. The guard was now standing there in his rubber suit. I asked him, "Where is everybody?" He replied, "They've all been told to assemble outside, next to Admin-2." I looked out of the window and, sure enough, there were our lads. Semikopov was there, and Simonenko, and the women from the chemical center. A few of the turbine people, too.

C: Were they just standing out in the open?

G: Yes, out in the open. Where the crossing used to be – to Block 3. Down below, on the right, I went out with my nose and mouth covered by the mask. I went over to my mates. They were smoking. I didn't smoke. Then I started to feel sick. I took my mask off and vomited. I was sick three times. Semikopov said, "You've bought it, lad." He took me by the arm and led me away. Next to Admin-2 there were three ambulances. Doctor Byelokon was on the telephone in the entrance to Admin-2. We went up to the doctor and Volodya said, "This lad is vomiting." Three drivers were sitting there. Byelokon said to one of them, "Take this man to the medical center." They put me in the ambulance and we set off. When we passed the turbine hall of Block 3 there were five engines already standing there. They stopped us and put three firemen in with me. I found out their names later: Kibenok, Tishura, and Titynok. (They all died subsequently.) Then we set off again.

C: Roughly what time was this?

G: It was just after 2 a.m. When we got to the medical center and they had filled in a card on me, I looked at the clock, it was 3 a.m. On the way to the medical center my nausea had subsided and I had got a grip on myself. But, seeing the way the firemen were vomiting and suffering, I thought to myself, "Well, I've caught some myself – but God help these lads!" When we got to the medical center they undressed all four of us. I had a wash and they put some pajamas on me. They filled in a card with my details and took me up to one of the upper floors of the Pripyat hospital. There they immediately put me on a drip. I had diarrhea and I vomited again, several times. They took off my watch and wedding ring and put them in a plastic bag, wrote my name on the label, and put the bag in a safe. Then they took me to the ward. They put me on another drip, but the problem was, I was being sick. I said, "Give me a bowl", but the nurse answered, "We haven't got enough bowls, just vomit on the floor." So there I was, the drip was putting the fluid in, and I was vomiting it out onto the floor. All this happened in the night. They poured three bottles of solution into me – into each of us. I did actually feel a bit better and went to sleep."

2.2 The Next Shift on Duty

We leave Oleg Genrikh for a while. In 1987, I was with Oleg and many other men who had been on duty in Block 4 on the night of April 26. We all received medical treatment in Clinic No.6 in Moscow. Of the seven people with whom Oleg had been in the medical center in Pripyat, five would die in Moscow. Tolya Kurguz died May 12, 1986. The doctors established that, apart from his serious radiation burns, he had received a dose of 12 Gy.
While the doctors in Pripyat were fighting to save Oleg Genrikh and many other men who had been brought in from the power station, Viktor Smagin, shift supervisor in Block 4 from the time of its first trials right up until April 26, 1986, was getting ready for work.[2]

C: Viktor, how did the situation develop in Block 4 and in the power station after the explosion?

S: The situation developed as follows. On the 26th I was due to go on duty at 8 a.m.. I got up as usual at 6 a.m.. I had just had breakfast when somebody rang my doorbell. I opened the door and there was my neighbor Yurii Byelozertsev, who is now dead. He told me that he had had a telephone call from the power

---

[2] This interview between the author (C) and V. Smagin (S) was conducted in Moscow in April 1990.

station, warning him to close all windows (because of high radiation levels). It was his friend, who had been involved in the experiments that night.[3] Actually, this was Zabolotnykh, the head of the laboratory in the chemical center. They had done some tests and found out that there was reactor fuel in the air. By the way, they had reported this to the power station management, too. After making that report, Zabolotnykh had rung Byelozertsev and said, "Shut your windows – there's been an accident and fuel is getting out into the air. The Devil only knows what's going on." So Byelozertsev, knowing that I was leaving to go on duty, came and warned me.

After that I went out onto our balcony and looked over towards the block. The central hall of Block 4 had been demolished. I lived on the 12th floor of a 16-storey complex in the center of Pripyat. My balcony looked out over the square, that is, it faced towards Block 4.

C: Could you see the fire?

S: It depends on what you mean by fire. There was smoke over the block. The sun had already risen. In the dark you could have seen the flames, but not in the daylight. You could see the ruined central hall – well, the whole structure was wrecked. I couldn't see what it was like inside, of course. A sort of stereotyped interpretation came to me immediately. I knew that experiments were due to be carried out in the daytime on April 25. Besides that, on the 25th I was not at work – the 24th and 25th had been my days off. On the morning of the 25th, I met Sasha Akimov on his way home from the night shift. I asked him, "How are things up there?" I knew there were going to be experiments to find out what power the turbines could produce when spinning to a halt after being disconnected from the steam supply, and after these running-down experiments, another experiment was planned - this was a cooling experiment - air cooling. That was the planned schedule.

C: Hadn't there been any running-down experiments (see footnote) before?

S: We did try it once in Block 3, but it wasn't successful. This was the second attempt. When I met Akimov on the 25th I asked him, "So, have they shut down the reactor for the experiment?" "No," he said, "when I handed over the reactor to Kazachkov, we were producing 50% of full power. The first shift is on now and they should be doing the run-down now – then, later, they'll be doing the air-cooling test."

This is why I already had an interpretation ready when I ran out onto the balcony and saw the damage. I had already realized that there had been some kind of explosion. Exactly what kind, nobody had told me. I had this fixed idea about what must have happened. The reactor was supposed to have been shut down. But they had obviously blown a separator drum when they were

---

[3] Other details of the experiments are given in Chapter 3.

doing the air-cooling test. It must have happened when they stopped pumping the cooling water.

C: You didn't yet know that the core had been destroyed?

S: No, of course not.

C: Did you even think that might have happened?

S: All I knew was that there was fuel in the air. But how much, I had no idea, of course. I could only assume that some fuel rod seals had blown and that the volatile part of the fuel had escaped into the atmosphere. That was my fixed idea. I thought, "A fine bit of cooling, that was!" It never entered my head that the reactor could have been working. After that, I closed all the windows. I woke up my wife – it was Saturday and her day off. The children were still asleep and soon they would have had to get up as it was a school day. I woke my wife and told her there had been an accident in Block 4. A serious accident. The air wasn't safe to breathe outside. I told her not to let the children out – that they should all stay at home. The children were not to go to school. I said we would decide what to do when I came off duty. I went to the bus-stop, it was right next to our building. At 7 a.m. I caught the bus to work, but I went to Admin-1 instead of Admin-2 as usual. The buses to Admin-2 had already stopped running and I imagined that they would be getting into Block 4 via Admin-1. When I got there they were not letting anyone in. Most of the staff were standing around in the entrance hall and only top management were being allowed in. As I am a block supervisor, I was included in that category and I had what's called a go-anywhere pass, so I was let in. When I got into Admin-1 I immediately asked where the managers were. I was told they were down in the bomb shelter. The entrance to the shelter was nearby, I went there. At the entrance I ran into two deputy directors, Tsarenko, deputy personnel director, and Vasilii Ivanovich Gundar, deputy general director. They were just coming out of the shelter. The first thing that Gundar said was, "Vitya, are you going on duty?" I said I was. Then he said, "Get changed in the glasshouse. There are suits in there." He meant our conference hall on the third floor. "Get a suit on and hurry over to Block 4. They've sorted out everything that they can sort out over there. Babichev is in charge there now instead of Akimov – you go and relieve Babichev."

Those were my orders. So, naturally, I got suited up and dashed over to Block 4. On the way, I called in at the Phase 2 safety equipment store and got myself a face-mask. I put on the mask and waders because I had to go through the damaged area on level 10. Water was pouring into the corridor along which I had to go to get to the control room of Block 4. It was a grim sight – awesome, in fact.

C: Were the dose monitors at work? Were you told what the radiation levels were?

S: There were monitors posted at the Block 4 end of level 10. They told me that nobody was allowed in. As I was going on duty, I just had to explain to these chaps that I was obliged to go in, and I went in. I got a warning at least, but no real indication of what the radiation levels were like.

C: Didn't they say why?

S: They just didn't know. When I got into the control room there was Babichev acting as shift supervisor. My former senior reactor engineer (SRE) Gashimov and the SRE who was then working with me, Alyosha Breus, were already there. These two SREs plus Borya Stolyarchuk, the SRE from the night shift. And the deputy chief scientific officer, Mikhail Alekseyevich Lyutov, was also sitting there at the control panel. So was the senior reactor engineer from my shift, Sasha Cheranev. Apparently, Gashimov and Cheranev had been brought in at about 5 a.m.. I don't know the exact time but they were brought in before the shift started.

The first person I ran into when I went in was Stolyarchuk, the first thing I asked was, "Was there water in the system or not?", but I repeat, I still had this fixed idea that the separator drums had burst. I asked, "When did the water run out of the core? What do the instruments show?" They responded, "Straight after the explosion." "What time was the explosion?", I asked. "Some time after 1 a.m.", they answered. Then I understood – although I had not yet grasped the extent of the damage. You couldn't see much on the way to Control Room 4 – only that the roof of the central hall had been blown off.

C: You couldn't see much from the transformator station either. From the Spent Fuel Store and the Liquid Waste Store there was a good view of the damage.

S: That's right. Anyway, at this moment Sichnikov turned up in the control room. Then Akimov came down with his team. They tried to open the valves. At first they had tried to turn on the emergency cooling system (ECS) – which is next to the emergency feed pumps, next to the deaeration unit. But they could not get into the ECS cubicle on level 27. So they tried to turn on at least the feeder system – the feeder network had apparently been turned off just before the explosion. The separator drums had been oversupplied so the senior reactor engineer had closed off the feeders. He had even blocked off the main feeders with gate valves. Akimov took his team and tried to get some water into the core, not knowing that by this time there was no core left.

C: So the plan was to get some water into the core?

S: There were two attempts. Akimov's men tried first. They turned on the start-up feeder network, these are small pipes compared with the main ones. There were five men in this team: Akimov himself, Toptunov, Sasha Nekhayev (who was sent from Phase 1 to help out), Slava Orlov, the deputy supervisor from reactor hall 1,[4] and Arkadii Uskov, the senior operating engineer also from reactor hall 1.

C: Arkadii was not supposed to be on duty. He came in that morning, though.

S: They all came because of the explosion. Just as Chugunov did. He was the supervisor in reactor hall 1. Orlov was his deputy. Arkadii Uskov was his senior engineer. Akimov took them up to open the gate valves because the order had been given to get some water in at all costs. That was the order constantly coming down from the top, "Get that water in!"

They had already found out that the emergency cooling system was out of commission. They had reported that the pipes were tangled in knots. I didn't go in there. I didn't see it. But they already knew by then. They couldn't get into the cubicle housing the emergency cooling and control system (ECCS) – that area was full of scalding steam and water. That's why they tried the feeder system and started to open the valves on the start-up network. On the side nearest the turbine hall stood Mekhayev, Akimov, and Toptunov while on the far side stood Orlov and Arkadii Uskov. Those on the near side got a horrific dose, but on the far side the dose was considerably less. On the near side there were ruptured steam pipes that were leaking onto them. That's how Sasha Mekhayev got his burns. Akimov and Toptunov, of course, also got a massive dose. Akimov, though, was not only there – he was running around everywhere and naturally got a big dose. They came down to the control room about 20 minutes after I got there. During that time I had just been running around looking at the instruments that had stopped working. I was trying to find out what state everything was in. I saw that the control and protection rods were inserted only 2–3 meters into the core. The rod indicators had stopped at that position. When I arrived there was already no water in the deaerators. And, as I was told in the turbine hall, the delivery pipe to the feed pumps was ruptured. But the turbine condensers were not damaged. When I arrived the condensers were starting to fill up with water – that is, chemically purified water was being fed into the condensers from Phase 1. They were feeding in this water so that later it could be released into the deaerators, but to do this it was necessary to switch off the second electric feed pump. So, in fact, there was a possible way of delivering it using the emergency feed pumps via the emergency cooling system. Everybody's thoughts were focused on this idea to the exclusion of all else.

When I arrived on the spot and reported for duty I was told that the main need was to get the water in. Naturally, I started working towards this aim, too. Then Akimov arrived and said, "Sorry, lads, we couldn't get the emergency cooling system (ECCS) working. We've got this start-up network going." What

---

[4] Reactornyi tsekh 1 – RTs-1

he had to show us on the diagram he could have shown us in 3 or 4 minutes. But it took him three attempts before he could explain it all because he was vomiting. He had to keep on running to the bin in the corner. He and Toptunov were there all the time. Mekhayev was in a bad way, too. He would sort himself out as well as he could, the spasms would stop and then he would come back and go on pointing at the diagram. He said, "When we went up from there I ran to this cubicle where the feed pump valves for the emergency cooling system are. The steam had stopped pouring into the cubicle – you can get in there now!" But looking at the others it was obvious that they were already in no state to try again. Sichnikov also said that he had been in this cubicle. He had opened one gate-valve by the access hatch, I think. All these gate-valves are electrically operated, but there was no power supply. Everything had been ripped to pieces. You could only open them manually. So Babichev, naturally, said, "Go on, take a group and get up there. Try to open them." I led the group that went back up. Arkadii Uskov and Slava Orlov went with me – it was their second time. The fourth man was Aleksei Breus, the senior reactor engineer from my shift, the shift that was supposed to be on duty that morning. The four of us went up there. We got to the cubicle. It was true that we could now get in, although there was water up to the bottom of the access hatch and that was quite a height from the floor. But three of us had waders on. Uskov, Orlov, and myself were in waders. Breus was wearing ordinary boots, so he climbed up on a pipe and started turning the valves by the hatch. We went to the far corner of the cubicle. First we opened the valves there, then we went back to Breus and finished off opening the valves that he had not got around to. We set up a delivery line to take water from the emergency pumps to the core. Then we went back down. On the way down, naturally, we realized that we were exposed to powerful radiation – especially as we were running down the staircase by the block wall. The wall was bare. All the glass had been blown out. We could see a lot of stuff lying around down below. We couldn't make out whether or not there was fuel down there. We could tell that the radiation level was high by the characteristic metallic taste in the mouth.

C: Didn't you have any dosimetric instruments?

S: We didn't have any instruments. Nothing at all. No. When we got back down a question arose. I said to Babichev, "Let me relieve you. I'll take your shift and you can go." Babichev responded, "I am still fit to work. I don't feel any need to go." I tried to sort a few things out and while I was doing this Lyosha Breus said, "When we were running downstairs I looked out of the window. At the far corner of the reactor section there's another staircase of the same type. There was water pouring down it." We thought for a bit. Where could that water be coming from? The only explanation was that it was from the fire extinguishing system. It could not be anything else. Especially as it was pouring in from above. We could see this better from the reserve control panel. The key is kept by the shift supervisor. I got the keys and Orlov and I went there. The reserve control room windows were all out and we could see all the devastation clearly – all the western side facing the spent fuel store. First, I looked at the water. It was, as we thought, a fire hydrant that had been smashed. Water was pouring out of it and the fire system pumps were banging away. They had not been stopped. Moreover, nobody would have dared to go and stop them – although nobody was using this water and it was just making the situation worse. I must emphasize this point – that according to the rules that were in force at the station even the shift supervisor did not have the right to turn off the water, because the fire pumps were the responsibility of the shift supervisor in the electrical section. But even he could not turn them off without authorization from the block shift supervisors and the station shift supervisor (SSS). And there was no real way of getting through to the SSS. There was no direct line. There were numbers you could ring, but just try ringing the SSS in a situation like that.

The water was my main concern. I saw that it was coming from the fire hydrant and that it had to be stopped as quickly as possible. But these fire pumps supply more than one block. They deliver water to two blocks. That's why one block shift supervisor alone did not have the right to turn them off.

What's more, the pressure in the fire system had to be maintained. If the fire pumps in Block 4 were stopped there would be no water for the system in Block 3. It was possible to break the connection between the two blocks, but that was no easy matter. It involved going outside and climbing down into the manholes (there were more than one). That's what I was contemplating. But Slava Orlov was looking at the devastation. He said, "Listen, Vitya, we're really in a mess. Look – there's graphite out there." He was right. There were graphite blocks lying outside. If the graphite had been blown out, then the whole core had gone.

C: So that's when you realized the scale of the destruction?

S: Yes, then we knew. Firstly, we had been wasting our time on jobs that would not get us anywhere. Secondly, a completely different disaster scenario opened up before us.

We went back to the control room. When I had been there before at the start of the shift the deputy chief scientific engineer, Lyutov, had been sitting there. He had said "Smagin, give me hand, I'm trying to find out the temperature of the graphite." I had said to him then, "How can you find out the temperature of the graphite if the power's been cut off by the accident? If there's no power to the thermocouples, how can you find out the graphite temperature?"

Now that Orlov and I had seen the damage, naturally, we were in a state of shock. We went back to the control room. There was Lyutov still sitting there. I said to him, "Mikhail Alekseyevich, you were asking me about the temperature of the graphite? Now I can tell you exactly."

"What is it?" he said.

I replied, "The temperature they gave out on the weather forecast this morning."

"What do you mean?" he asked.

"Just what I said – the air temperature. The graphite is lying outside."

"It can't be!" he said.

"Come with me", I said.

We went back together with Uskov and Lyutov. Orlov asked Lyutov, "Do you see those blocks lying there?" Lyutov answered, "It is probably a reject rod assembly." Orlov swore at him and said, "You can see there's a male connector on it. When did you ever see a rod assembly with a male connector?"

With that Orlov said to Uskov, "There's nothing we can do here. Let's go". And they went. We all went back to the control room. It was then that an order came, "Babichev, please come immediately and report on the situation." They needed a competent person, a technical expert, who knew the system in Block 4. Babichev was the phase shift supervisor. He knew Block 4 perfectly well.

Then another order: "Smagin, take over the shift. Babichev, report to the management in the bomb shelter."

Taking over the shift involved making a tour of inspection. I had not yet been in the turbine hall. Babichev warned me that the turbine hall was dirty. But I couldn't take over the shift without inspecting everything. So I had to run in there and take a look. I went up to level 12. The devastation was appalling. I went back down. I tried to ring the station shift supervisor. I reported that water from the fire system was flooding the block. Soon the whole bottom floor would be flooded – and the lower levels interconnect with the other blocks. Had Block 3 been shut down? Yes, it had. During the night, at about 5 a.m..

So, I reported to the station shift supervisor that the fire hydrants had been smashed and that water was gushing out. They had to be cut off so that Block 3 would not be left without water to fight fire with. The station shift supervisor gave the necessary order, of course. The electricians went and did the job.

Before Babichev left, a dosimetrist from my shift, a certain Nepyushchii, had run around doing a check. He ran past like a meteor shouting, "800, 1000 microRoentgens per second!" I stopped him, "Let's sort out this radiation level, shall we?! It turned out that his instrument had gone off scale. So, nobody had any idea what the radiation level really was. When I realized that we were getting a very high dose I cleared all the staff out of the control room. Some I sent to Block 2. I told my senior reactor engineer, Breus, to go to Block 3. "Stay there", I said, "if I need you, I'll call you". I had a bit of a tussle with Breus. He didn't want to go.

C: Why?

S: That was the attitude of the staff. You work to the end, whatever the circumstances. The orders kept coming, "Get that water in!". The idea was to prevent the fuel in the core from melting. After I had sent the shift away I started vomiting.

C: How long were you in the block?

S: Time seemed to stand still. By 2 p.m. the electricians had reported that they couldn't get the condenser pumps working. It was then that I realized there was no point in staying in Block 4. I rang the station shift supervisor (SSS) and reported the situation. I was ordered to move to Block 3.

I went to Block 3. My senior reactor engineer (SRE), Breus, was there and Usenko, head of the turbine shop, who had also had a big dose, and Kovalev, senior machine operator in the turbine shop.

We were all running in turn to the bin. There was nothing left to vomit up. Yellow bile was coming out. We were starting our second shift of the day, having started at 8 a.m., April 26. In the night they had taken the whole shift to the hospital. I got through again to the station shift supervisor and informed him, "These lads have had way over their dose." He gave the order, "Get to the sick bay quickly."

C: Did you realize that you had had a big dose?

S: Yes, in the radiation safety instructions that are followed in every nuclear power station it says that vomiting is the first sign of radiation sickness. Lesha Breus didn't feel quite so bad. So we left him in the control room in Block 3, while we went off to the sick bay.

C: Who else was with you?

S: My senior reactor engineers Gashimov and Cheranev, Usenko and Kovalev. We got to the sick bay. Imagine the scene. The doctor was trying to fill in our details. Then one of us saw something that looked like a bucket and made for it. The problem was we all got the spasms together. As soon as one started, so did all the rest. We all emptied ourselves into that bucket. The doctor couldn't even get our cards filled in.

Then they got us into an ambulance and took us to the medical center in Pripyat. They put me on a drip immediately.

We lay there all that night. We heard the weird sound of the safety valves blowing on Phase 1. They were shutting down Blocks 1 and 2. They shut them down in the night to April 27. Block 3 had been shut down in the early hours of the 26th.

C: When did they send the first group to Clinic No.6 in Moscow?

S: That was at 10 p.m. on the evening of the 26th. There were 27 people in that group. There were two on stretchers in the ambulance – Kurguz and Degtyrenko. Shushenok died at 12 noon on April 26.

C: What had happened to him?

S: He had very bad burns.

C: Was he on the staff at Chernobyl?

S: No, he was a visiting engineer. He was there for some trials. He had been in a cubicle under some steam pipes. There were some gauges attached to these pipes. He was there next to these instruments. When the pipes burst he was badly scalded by the hot water. He was dragged out by Palamarchuk, Shushenko's laboratory chief, and Kolya Gorbanenko.

C: And when did they send you to Moscow?

S: At midday on April 27, by special flight.

C: What was your reception like?

S: They drove us around Vnukovo airport for about an hour, not knowing where to put us down. We were "dirty", you see. It was an appalling scene. The doctors that came to meet us were dressed in protective suits and aprons made of PVC. The seats on the buses were covered with polythene. We were in all sorts of gear. We didn't have enough clothing.

C: How did they select you for this special flight? What was the criterion?

S: Burnt skin is the sure sign of a big dose. That's how they picked us.

C: How many of you were there?

S: Twenty-seven in the first party, 130 in the second. Then they brought a third group and even a fourth. That's as much as we saw.

C: Which part of the hospital were you in?

S: On the 4th floor. I was with Chugunov, Dyatlov, and Uskov. While they were still allowing us to walk about Sasha Akimov used to come and see us. But he died on the 11th of May. Then it really got going. The lads started dying.

C: When did they work out your dose and tell you how much you had had?

S: They didn't tell us our dose. I found out my dose by accident. When the doctor was doing his rounds. We were going through the crisis. They took marrow samples for analysis. The doctor mentioned my dose as he passed on his rounds – it was 2.7 Gy. He deliberately used Gray units, so that I wouldn't understand.

C: What sort of doses had the other lads received?

S: Those who officially had radiation sickness, I can list for you: Sasha Mekhayev, third degree radiation sickness, 5 to 6 Gy; Yura Tregub, second degree severe radiation sickness (SRS), 4 Gy; Vladimir Aleksandrovich Chugunov, former supervisor of reactor hall No.1, second degree SRS; Slava

Orlov, Chugunov's deputy, second degree SRS; Arkadii Uskov, second degree SRS; Telyatnikov, second degree SRS."

2.3 Afterwards

Now I return to Oleg Genrikh.[5] *I think that the following reflections of his provide an appropriate closing to this chapter. Not everybody will appreciate his comments and thoughts, yet, I have strong feelings that they should be included here.*
Genrikh went through the horror of the accident, and through subsequent complications, physical torment, and treatment. He received such a large radiation dose that it is inconceivable that a human being could then survive for such a long time thereafter. The general belief is that out of a group of individuals that had received a dose of about 5 Gy, 50% would die within a couple of days and the rest a bit later (see also Appendix A).

C: Oleg, if you had the opportunity to appeal to the Soviet authorities, to the Supreme Soviet on the one hand and, on the other hand, to the international community, what would you say to them? What help would you hope to get? What would you like to see changed in the way Chernobyl veterans are treated, those people who suffered from the accident itself or ruined their health during the rectification work?

G: To the Supreme Soviet, I would address the request: Give us legal status, recognize us legally!

That would make the problems easier to solve. Because, whoever you turn to, you just have to utter the word "Chernobyl" and they all try to get rid of you. I don't understand. We've got glasnost and perestroika. Do these mean different things or the same thing? Or is this just for the television and for the in-crowd? When will they [the government] come to grips with the social and medical problems? That concerns also the guys who are in hospital now. I was in hospital with a miner who built a tunnel under the foundations of Block 4 at Chernobyl. He went downstairs to telephone his wife and he couldn't get back up. I went down on the elevator and brought him back up.

Give us legal status like the Afghan veterans. I don't want to ask for more than the Afghan veterans get. I can't judge the contribution they made. I have just this one request – legal status. If we get that, then when we go into any bureaucrat's office, he will have to see us and do something about our problems. And it won't be him pressuring us, it will be the other way around, because the law will be on our side. Just now, everyone we turn to has the same excuse,

---

[5] This is a continuation of the interview that begins in Section 2.1.

"The law says nothing about such cases. If I do this for you, I'll be in real trouble." They're all afraid of being in real trouble. Well, we've already been in real trouble, as real as you can get.

As for the international community, I would wish that nobody out there ever lands in the situation my mates and I are in. Don't let such accidents and mistakes happen – although ordinary people can't do much about it. I hope to God nothing like Chernobyl ever happens anywhere else. And you should help those who have suffered. In America there are Armenian Societies. In Canada there are Ukrainian Societies.

My name is Genrikh – Heinrich. I am a German. I have a lot of relatives in the Federal Republic of Germany and I make no secret of that. Why couldn't a Frenchman help us, if he's got the money? If I want to help my friends I do what I can for them. I buy flowers for my doctor, for example. It isn't a bribe, honestly. It's just a token of appreciation. I also light candles for people in church. I don't take any money for that from the trade union, I pay for them myself. More than that I cannot do – I live on a pension. But what about those people who live in other countries, have businesses and money? Help us, please. If only with words on our behalf. If it is material help in the form of hard currency – then thanks again. If you could send medicines, disposable syringes – by all means. Thanks a lot. This would be pure generosity. It isn't that we are begging, that would be wrong.

I have grown more compassionate as a result of all this. It's terrible how people snarl at each other nowadays. We need to do quite the opposite – we should unite and help each other. Unity is what we need, not fragmentation. I watch the television – when the patriarchs and the priests call on us to be compassionate, they're right. When a person finds himself in a situation like that, even if he is not on the edge of life and death, he becomes more compassionate, more merciful. I consider that I was on that boundary. Such a person becomes more merciful towards others, and towards what is going on in the country, and altogether towards everything that lives on the earth. And that's how it should be. The Church is holy, a human being needs to have something holy. For some it may be a political party, comical though that sounds. Let them have it. For someone else again it might be his wife that's the holy person, for another it might be his mother.

Every human being needs a faith. You can't live without faith – only animals can do that. But when you have faith then you truly have a sense of proportion, a sense of tact. You have a future – and this future is possible, it will come, as long as we never forget about Chernobyl.

Nothing like it must ever happen again, anywhere."

3. Who is Really to Blame? – Designer or Staff?

"You must understand that the reactor does have some flaws. It was designed many years ago by Academician Doldezhal, who had only the technology of that time available to him. Now these flaws have been compensated for.

"The problem isn't the design. If you're driving a car and turn the wheel in the wrong direction and have an accident, do you say that the engine is at fault? Or its designer? No. Everyone will say that it was the fault of the driver..."

Academician Anatolii P. Aleksandrov
Letters to the Editor
"Ogonyok", No. 35, August 1990, p. 7

Five years have past since the accident and many experts are still thwarted in their attempts to get objective information on what occurred at the Chernobyl Nuclear Power Station, especially, on the night of the accident and on the following day. More than half a decade later and they still lack details that would enable them to make an accurate assessment of the causative circumstances, technology, and organization. Much has been written since the accident.[1] Nevertheless, most people have not yet looked beyond the official explanation (to the IAEA) that *"the main cause of the accident was a freak combination of infringements of rules and working practices on the part of the reactor staff"*.

The February 1990 decision of the scientific and technological section of the USSR National Nuclear Safety Committee (GosAtomEnergoNadzor), which oversees nuclear safety at operating power plants, stated that – because of the absence of any

---

Preceding page, an operator at the control panel of the RBMK reactor; those of Blocks 3 and 4 were identical.

[1] For further material see Section 1.3, Appendixes B, C, and references [3.1 – 18].

indications of possibly dangerous reactor configurations in the
RBMK design – the accident was caused by personnel untrained
in safe operating conditions who violated regulations.

Yet, for a reactor with an adequate emergency system, the
safety violations and errors committed by the staff could have
and should have led to an emergency shutdown of the block –
the last actions taken by the staff were in fact directed towards
shutting down. And it was precisely these actions that, because
of the design specifics of the reactor and the lack of fail-safe
mechanisms, led to a catastrophic growth in power and the subsequent destruction. Yes, of course, it was the staff who caused
the explosion, but it had, in effect, been designed to explode.

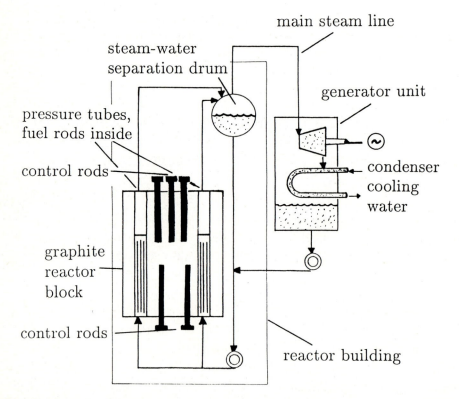

Fig. 3.1. Schematic representation of the Chernobyl graphite-moderated reactor. To control the
reactor, bundles of rods are inserted into pressure tubes placed in the graphite moderator block.
The fission heat converts the water rising through the tubes to steam. The steam drives the turbine
and is then reinjected to exploit the remaining heat. There is no overall containment structure, but
the core is simply confined by a steel mantle (with a layer of inert gas in it)

3.1 The RBMK: Design Regulations and Safety System

Perhaps it is worth taking another look into the matter. In this complex under-taking we shall need the help of experts whose work is not known beyond a small circle of their fellow professionals. We shall also need to examine the documents and regulations on the safety of nuclear power stations which are to be adhered to by all organisations and institutions taking part in the design, construction and operation of nuclear power plants. These guidelines are to be found in the *"Safety Regulations for Nuclear Power Stations"* (SRNPS-04-74, below referred to as "Safety Regulations") and the *"General Safety Guide-lines for the Planning, Construction and Operation of Nuclear Power Stations"* (GSG-73, below referred to as "Safety Guidelines") [2] It should be noted that both sets of regulations came into force before work began on the construction of Chernobyl's Block 4.

Returning to the official explanation of the accident, we might compare it with the Institute of Atomic Energy in Moscow (IAE) analyses *"Investigation into the Causes of the Accident at the Chernobyl Nuclear Power Station"*, which was published in October 1986 with the approval of the deputy director of the institute Ye. P. Ryazantsev (D.Sc. in technical sciences), one of the experts involved in the preparation of the report to the International Atomic Energy Agency (IAEA) in Vienna.

The Institute of Atomic Energy report concludes that *"... the main cause of the accident was a freak combination of infringements of rules and working practices on the part of the reactor staff, under which faults in the design of the reactor and its automatic control and safety systems became apparent ... "*.

Let us now consider the implications of this statement in more detail.

The part of the Institute of Atomic Energy's conclusion following the words *"on the part of the reactor staff"* did not find its way to the IAEA; it was omitted.

What were these "design faults" which the scientific advisor to the Com-mission regarded as of so little importance that they are only included in the Institute of Atomic Energy's own (internal) report as items classified "for in-ternal use only"?

---

[2] The Russian abbreviations are PBYa-04-74 and OPB-73. The last numbers denote the years in which the regulations were approved.

3.1.1 "Unimportant" Design Faults?

Professor B. G. Dubovskii is one of the authors of the SRNPS-04-74 regulations, and for 14 years (1958–73) he was the head of the USSR Nuclear Safety Board, which was set up in 1958 on the initiative of Academician I. B. Kurchatov. In his 1988 paper on the instability of the RBMK reactor, he writes:

"It is beyond comprehension how those in charge of the design of the control and safety systems for the RBMK could make such serious miscalculations, some of which defy the most elementary logic. It is no exaggeration to say that the RBMK reactors had no proper safety systems until 1986. They had no accident protection whatsoever! – Neither below the core, nor above it."

Another expert, A. A. Yadrikhinskii, safety inspector at the Kursk nuclear power station, concludes his analysis *"The Accident in Block 4 of the Chernobyl Nuclear Power Station"* thusly:

"The nuclear accident – the explosion of Block 4 of the Chernobyl Nuclear Power Station – was the result of infringements of the SRNPS-04-74 regulations by the chief scientific officer and chief designer."

He lists seven requirements laid down in these rules. However, a thorough analysis of the RBMK design as given in the same work has revealed 32 infringements of the SRNPS and GSG regulations.

In the 1987 official article *"The Foreign Press and the Chernobyl Disaster"*, the following observation is made:

"In complete contrast with the official Soviet verdict, British experts believe that design faults, not human error, were the primary cause of the disaster."

The British regard the mistakes of the operators as *"contributing factors"*.

The same source gives the opinion of American experts that *"The reactor's trip mechanism itself was the cause for the sudden (power) surge."*.

In the above-mentioned Institute of Atomic Energy report 13 possible versions of events leading up to the explosion, including sabotage, are considered. In the end the writers conclude: *"It is sufficiently obvious that the only version which does not contradict the available data, is the version related to the effect of the displacer rods of the automatic control and protection system."*

For the non-specialist it is necessary to add a short remark: the disastrous effect of the faulty control and protection systems on the reactor is due to a deficiency in their design – at a core height of the RBMK of 7 meters, the length of the water displacers was chosen to be only 4.5 meters.

Many people, even the designers of the RBMK, admit that there are fundamental deficiencies in the design of the control and protection systems (the so-called emergency system of the reactor).

Are there really any deficiencies in the construction of the RBMK reactors?

Let us refer to one of the basic requirements for control and protection systems in the SRNPS rules (SRNPS §3.3.26):
"The safety system of the reactor has to ensure a fast and reliable automatic shutdown of the chain reaction in the following circumstances:

- *if the power output reaches a dangerous level;*
- *if there is a dangerous power surge;*
- *if technical faults develop which make it necessary to trip the reactor;*
- *if the emergency button is pressed."*

It is worth emphasizing the statement that the safety system has to ensure *"... a fast and reliable shutdown of the chain reaction ... "*. There should be no prevarication about a reduction in the effectiveness of the emergency systems such as that which provided a refuge for the writers of the report to the IAEA: *"In the situation created, the infringements of normal practices by the staff led to a substantial reduction in the effectiveness of the emergency safeguards"*. A reactor, after all, is not a car. No fault in its "braking system" can be allowed, whatever the circumstances. This is clearly spelled out in the regulations (SRNPS §3.3.28):
"... in the design of the reactor it must be shown that in ANY CRITICAL SITUATIONS the active parts of the emergency systems – even if one of its parts is not functioning – ensure ... a fast and reliable shutdown of the chain reaction without damage to the fuel rods."

Once again, we need to emphasize the wording of the regulations: *"in any critical situations"* and *"without damage to the fuel rods"*, not to mention an explosion which hurls the fuel (of which the RBMK core contains about 190 tons) into the environment. This is exactly what happened on April 26, 1986.

In order that a reactor's emergency systems should always perform their function correctly in any critical situations, designers are required by the regulations (SRNPS §3.3.27) to include in the design *"no less than two independent groups of emergency protection mechanisms"* – in other words, at least two fast-acting "braking systems", in case one of them should fail.

3.1.2 The RBMK's Safety System

"The explosion occurred after the emergency button had been pressed, which by itself looks quite paradoxical" as the scientists of the Institute of Atomic Energy comment calmly in their own report. However, this is only in their internal report, i.e., for themselves; when addressing the IAEA and the public, this fact is again concealed, or at least played down.

Another view of this situation is provided by V. I. Smutnyev, shift supervisor at the Novovoronezh nuclear power station, who states in his article *"What was the Main Cause?"* that *"... this is the sort of thing a nuclear reactor operator can only imagine in a nightmare ... "* He is absolutely right. Many colleagues share his opinion. Nevertheless, some scientists and nuclear experts obviously think otherwise – for example, the technical experts, who, when questioned in court about the safety of the RBMK reactor, replied: *"The RBMK is equipped with a reliable control and safety system capable of handling all regimes, stationary and transitional."*

3.1.2.1 Imagine: The Brakes of Your Car Accelerate!

It was the very emergency system which is supposed to shut the reactor down reliably and quickly, whatever its condition, that caused it to run away. (A steam-void effect may have been involved, too – on which we will say more below.) The explosion occurred 5 seconds after the emergency button was pressed. Three seconds after the button was pressed, when the power level was at 520 MW, the emergency alarms were activated by the power rise from the preceding level of 200 MW and by the sharpness of the rise (see above, SRNPS §3.3.26).

Thus, the accident prevention system failed to trip the reactor, not just in the normal situation, but even in this extreme situation.

How, in fact, could the safety system trip the reactor when at that very moment it was causing it to run away?

In their INTERNAL report, the writers of the Institute of Atomic Energy do not conceal this fact from themselves:

"With the reactor in the state that it was in, the pressing of the emergency button could, within the course of the first few seconds, lead to an increase in reactivity and run-away of the reactor."

In our opinion, to design a reactor with such an accident prevention system is equivalent to designing cars in which, in a moment of need (for example, on a steep descent) the brake pedal becomes an accelerator. Worse still is to keep quiet about this strange, or rather terrifying characteristic of the brake pedal and then – after a crash in which the wretched motorist attempted to stop his car trustfully using this "reliable" brake pedal – to accuse him of not understanding the braking system properly and of being reckless.

That was the kind of accusation leveled at the staff of the Chernobyl nuclear power station.

While there had been a lack of securing the safety of the RBMK design, there was no lack of authority in the repeated assurances of the designers concerning the safety of their reactor. Thus, the chief designer declared:

"The core of the RBMK reactor and the fuel pellets as well as the reactor control and protection systems and their active parts were all designed with the basic

safety requirements of nuclear power stations in mind – the safety of the reactor is assured in all working regimes and all states of the reactor, as well as in all possible emergency situations."

This statement is in flat contradiction to reality.

3.1.2.2 What Were the Designers of the RBMK Relying on?

We find the answer in the information supplied to the IAEA: *"The slow operation of the control and safety systems is compensated for by their number."* Let us now try to establish how, in this case, quantity could indeed "compensate" for quality.

The 211 control rods could be inserted into the core at a rate of about 40 cm per second. The height of the core is 7 meters. Thus, the time required for full insertion of the rods into the core averages 18–20 seconds.

Is 18–20 seconds a long time or a short time? A long time, a very long time.

Let us remember that the explosion took place in the fifth second after the activation of the emergency systems by the staff, i.e., after the pressing of the emergency button. But this was a terrific explosion: the power of the reactor at the moment of the explosion has been estimated by Institute of Atomic Energy scientists to have been 200 to 400 times the nominal power of the reactor of 3200 MW. In the third second the standard control system recorded the runaway of the reactor as being at a power output of 520 MW.

Note that, of the two groups of control and safety mechanisms mentioned above, only the second one, the group of emergency rods, is always in readiness, in the so-called "accident alert regime" in its place over the core.

Was this group of emergency rods capable of carrying out its function, as defined by SRNPS §3.3.28, to prevent the reactor from running away?

It would be capable of doing so, but only if its action were not accompanied by a rapid build-up of excess reactivity. In the case of such a build-up it was not in the least capable of doing so.

To give the non-specialist an idea of the speed of operation of the emergency rods we can again compare them to a car brake pedal which only applies the brakes after being pressed flat to the floor for 18–20 seconds. Would such a "fast-acting" braking system be effective in all possible eventualities? It is not difficult to answer this question. Thus, Professor B. G. Dubovskii states:

"An emergency system which takes 18–20 seconds to operate is not a protection system at all – it is a parody of such a system. Normal emergency systems, as used in reactors all over the world, come into operation in just a few seconds (up to 5 seconds at the most). At least, that is true of the rapid-acting subsystems."

Only after the accident of April 26, 1986 were the emergency rods modified to become a rapid-acting system taking 2 to 2.5 seconds to achieve full insertion.

Despite all the above, we read in the information supplied to the IAEA:

"The reactor's control and protection system ensures emergency shutdown of the chain reaction by means of emergency control rods actuated by dangerous deviations from the reactor's normal parameters or by equipment failure ..."

What is this, if not half-truth?

At this stage, the reader may well ask, "How is it, then, that none of you managed to blow yourselves up long before April 26, 1986?" A good question.

In order to answer it we have to consider the part played by the first group of control rods, the normal working rods, as opposed to the emergency group. What is its contribution to the reactor's emergency protection?

As has been explained above, the reactor staff controlled the power output of the reactor in all phases of its operation (start-up, operation at normal levels, planned shutdown for repairs) by means of the working group of control rods. In emergencies, this group would also be used to assist the group of emergency rods. As stated in the information given to the IAEA, their "increased number" should make up for the slow operation of the emergency rods.

For the first 7 years of RBMK operation, 1973–80 (the first block came on line at the Leningrad nuclear power station in 1973), there was no restriction at all on the number of rods of the working group which could be extracted from the core. More precisely, extraction of all rods from the core was permitted.

The number of rods inserted into the core during its operation is known as the operational surplus of reactivity (OSR). In other words, we can therefore state that there was no minimal permissible OSR value.

For this reason, it was by no means a rare event when a reactor of the Leningrad, Chernobyl, or Kursk stations (the RBMK reactors then operating) were started up and brought back to normal power generation after a brief shutdown with an OSR of only 1–3 rods. (This was the so-called "start-up from zero without passing through the iodine well".) There were quite a few of these short shutdowns, since the RBMK was just "starting out in life" and no-one had any experience in its operation.

All these years (and in the years that followed, right up to the accident) the staff of these stations was worried about just one thing – how to avoid big fluctuations of power output which might burn out the fuel rods and channels. That is, a heat transfer crisis caused by excessive thermal loading.

By 1980, it was standard practice at the RBMK stations to leave in the core about 30 of the original 200–240 supplementary moderators, installed to compensate for the extra reactivity of fresh fuel. Hence, in the 1980 technical regulations of the Chernobyl station is the first appearance of a restriction on the operational surplus reactivity (OSR): *"Operation with an OSR of less than 10 rods is not permitted."*

In the 1983 rules this restriction was tightened (by this time practically all the supplementary moderators had been replaced by fuel rod assemblies).

"If the OSR drops to 15 rods the reactor must be shut down immediately ... The scientific management of the station is required to carry out periodic surveys (once per year) of the actual stability of the energy output levels of a given reactor and, when necessary, to revise them in the direction of greater stringency in consultation with the Chief Science Oficer and Chief Designer."

There is no further explanation, not even a hint of the fact that an operational surplus reactivity value of 15 rods is the limit below which the reactor becomes dangerous (liable to explode) if the emergency rods are being lowered. We do not even find any eye-catching warning like:

ATTENTION! ATTENTION!
Infringement of this limit renders the reactor dangerous!

Admittedly, later on this nightmarish warning did appear in the technical regulations but, along with much else, only after the accident of April 26, 1986.

The figure of 15 rods, as well as that of 10, is introduced according to the above-quoted phrase from the technical regulation concerning maintenance of a stable energy output. The figure defines the limit of the regulatory capability of the working group of control rods. It does not lay down the way it should function in a protective capacity which, however, was how the staff of all RBMK reactors understood its function.

From the very first years of RBMK operation the station staff noticed anomalous power output behavior in cases when individual control rods of the working group were moved separately.

3.1.2.3 The Anomalous Behavior of the Safety System

When a rod which was at the extreme upper end above the core would be lowered into the core, the first meter of its insertion was sometimes accompanied by a brief surge (jump) in power output instead of the expected drop. The staff gave this phenomenon the name "rod-end effect" although "rod-end defect" would have been more precise. Science was later to call this "a positive surge of reactivity".

It is necessary to emphasize that such behavior was only observed in cases where individual rods situated above the core (at the top end switch) were brought into operation, and even then not in every case, but only with a certain shape of the energy generating neutron field.

Reports of this "strange" effect naturally reached the chief designer and the chief science officer. The essence of the problem was clear to practically everyone involved in the design: it lay in the faulty design of the control rod, in the wrongly chosen length of the rod itself and its attached component – the so-called graphite water displacer: While the height of the core is 7 meters, the length of the graphite displacer was chosen to be only 4.5 meters, and that of the absorbing section 6 meters.

When the rod is raised to its extreme upper position, the midpoint of the displacer is at the midpoint of the height of the core. It is easy to calculate that in this configuration there is a 1.25-meter column of water under the displacer and another such column above it.

The graphite displacer enhances the absorption capacity of the moderator rod since the neutron-absorbing capacity of graphite is significantly less than that of water. Thus, when a control rod moves down into the core from its top position, it displaces a 1.25-meter column of water in the lower part of the core, and introduces a positive reactivity into this region.

Later, soon after the accident, the core was redesigned so that the lower column of water was eliminated by the use of a shortened displacer and with a new top position for the rod, 1.2 meters lower than before. This was adopted as a temporary measure until the rods could be replaced by a new design (which took on the order of 3 more years).

Despite the strange reactivity behavior which accompanied rod insertion and which was observed even at the first trials of these reactors, they were approved fit for service by the government commission.

However, in February 1984, a letter arrived at the Chernobyl nuclear power station from the chief designer's office. It bore the signature of the director of SRDIET Yu. M. Cherkashov and the heading *"On the subject of positive reactivity surges"*. After an explanation of the basic features of *"the design characteristics of the RBMK control rods"* (not designer's errors, but design characteristics!) comes the following:
"We emphasize once more that positive reactivity surge will be observed only when the rods move down from their extreme upper position and only for a downwards distorted neutron field."

We note that at the crucial moment on April 26 when the operator pressed the emergency button, the neutron field was distorted UPWARDS, and not downwards. Yet, there was a surge.

Cherkashov's letter continued:
"Thus, we are dealing with a known phenomenon, concerning which decisions have already been taken... There are still further suggestions which lead to the following:

1) Do not use rods with short joining links for the automatic regulators. On all working reactors the displacers were removed from such rods.

2) Limit the number of rods withdrawn completely from the core (to the top end channels) to a total of 150. The remaining partly inserted rods must be inserted into the core to a depth of no less than 0.5 meters.

3) Carry out design modifications to lengthen the telescopic link between the displacer and the rod in order to displace the column of water from under the displacer. This will eliminate occurrence of positive reactivity surge for arbitrary deformations of the field.

4) The use of control rods without displacers and with special thin-film-cooling[3] *of the channels eliminates the possibility of a positive reactivity surge whatever the state of the reactor and for any configurations of the strong field of the power output. Control rods and channels of this type were envisaged in the technical design, but were rejected because of insufficient experimental testing. Sufficient experience with thin-film-cooling has now been accumulated so that a return to this scheme is justified.*
The tasks set out in the letter have been incorporated in the plan and work on them is in progress."

Let us evaluate these measures.

The measures described in items 3 and 4 of the letter were not carried out until 3 years after the accident.

Two years before the accident, as in item 2 of this letter, the chief designer recommended that the number of rods allowed to be extracted completely from the core should not exceed a total of 150. (The remaining, partially engaged control rods of the control and safety systems, if inserted not less than 0.5 meters into the core, would, in the designer's opinion, prevent a positive reactivity surge.)

In all this, Cherkashov said nothing about the position of the (additional) safety rods which are, in case of emergency, to be inserted into the core from below. (The reason is clear – the very thought of the formation of a local critical mass in the lower section of the core is inadmissible.) As we said above, these additional safety rods which should have been inserted from below did not enter the core, neither when the emergency alarm sounded, nor when the emergency button was pressed. As the accident showed, this was yet another glaring design flaw since there was no guarantee of protection for the lower part of the core – in fact, the protection system itself could create a local critical mass in that section.

According to calculations, the accident would not have happened if just this had been attended to before April 26, 1986.

What is a "good" minimal operational surplus reactivity (OSR)? Let us now evaluate the OSR value which results from compliance with the Chief Designer's recommendation.

The calculation is very simple. Take the 211 control rods. Subtract the 150 rods allowed to be extracted fully. That leaves 61 rods which can be partially inserted, to a minimal depth of 0.5 meters. The total length of the absorbent control rods in the core will then be 0.5 meters × 61 = 30.5 meters. The length of the absorbing section of a single rod was 6.2 meters, so the minimum number

---

[3] The inside of the rods is usually cooled by water. To avoid direct contact between the water and the inside walls of the rods, a thin film of another liquid is injected, thus, thin-film-cooling (plyonochnoe okhlazhdeniye).

of control rods – the operational surplus reactivity – which would comply with the recommendation for the avoidance of a positive reactivity surge would be 30.5 meters/6.2 meters = 5.

Not 15, as the designers of the RBMK now maintain. And what about the first series of reactors, those at Leningrad, Chernobyl, and Kursk, where there are only 179 rods? In those cases the same calculation gives the minimum number of rods as 2.5!

The scientists who analyzed the causes of the accident calculated that the operational surplus reactivity at the beginning of the experiment was 6–8 rods. Even laying aside doubts as to the correctness of this figure (according to the evidence of some of the people who took part in the experiment, the operational surplus reactivity was about 18 rods 30 minutes before the experiment started), it is more than the minimum recommended by the chief designer.

There is one more thing that has to be said concerning the protective function of the working group of control rods:
If the operational surplus reactivity value (whether it is 15 or any other figure – it does not matter in this context) determined the capacity of the working group of control rods in its protective role, then why the complete lack of any warning or emergency alarm system relating to this parameter? Why no visual display or recording apparatus? Why no protective blocking mechanism? The latter is one of the basic safety requirements laid down in the rules (SRNPS §§3.1.8, 3.3.21).

Besides, the operational surplus reactivity value has to be calculated, and the calculation takes 5–15 minutes. In order to find out this value, one has to use a computer, but the standard "Prisma" computer program, when working out the OSR at low levels of power output (up to 10% of nominal) is unreliable. Again, it was only after the accident that operational surplus reactivity display and recording devices were installed and that a full visual and acoustic (siren) warning and emergency alarm system was installed.

Why is there now a plan to install an automatic block which will trip the reactor when the OSR falls to 30 rods? – Now it is not to 15, but to 30! The scientific experts who were questioned by the court on this matter answered, in essence, that experience had shown the need to make the reactor "foolproof", to reduce its sensitivity to operator error.

Let us sum up all that has been said about the emergency protection system.

Before the accident, the RBMK reactors operated with a control and protection system consisting of two groups of rods, one of which was a caricature of a protection system (taking 18–20 seconds to come into operation instead of the required 2–3 seconds) while the other in certain specific conditions (for the minimal permissible operational surplus reactivity value?!), instead of protecting the reactor, could become, for the first 3–4 seconds, a device for causing a runaway reaction.

Only now does one understand how long we walked along the edge of the precipice.

It was good fortune that saved us all those years from a disaster that might have been called the Leningrad disaster, or the Kursk disaster, or indeed, as it turned out in reality, the Chernobyl disaster. And it was not the absence of a *"freak combination of infringements of rules and working practices on the part of the reactor staff"* which brought the reactor to a totally impermissible state considered unfathomable by the designers.

3.2 Economics – Technology – the Human Factor

How could the economical operation of the RBMK take precedence over its safety?

After all, in the original RBMK plans (1965) it was envisaged that each control rod would have an absorbing section and a full-length displacer of 7 meters. The 1969 technical drawings of the RBMK already showed groups of control rods with shortened absorbers (5 and 6 meters) and displacers (5 meters), but with thin-film-cooling of the channels, which eliminated the possibility of a positive surge of reactivity and, therefore, of an accident such as that which occurred at Chernobyl.

However, the working drawings of the RBMK showed control rods with shortened absorbers and displacers and with water-filled channels for the rods. Thin-film-cooling of the channels is more complicated to construct and, naturally, more expensive. But it would have eliminated the columns of water standing under the shortened displacers and, as a result, the positive reactivity surges upon insertion of control rods into the core. Moreover, it would have speeded up the operation of the emergency rods by eliminating the resistance of the water to the descending rod.

It took the Chernobyl disaster to bring about a return to the original plans and technical drawings of the control and protection mechanisms, and to secure installation of film-cooling in the emergency rod channels. Only after implementation of these changes could the group of emergency rods be justifiably referred to as a fast-acting accident prevention system (FAP).

The time taken for full insertion of the group was cut by nearly 90%, to 2–2.5 seconds. The length of the absorbing section of the control rods was increased to 7 meters. An increase in the length of the excluder to 7 meters is also planned, which will involve either the lengthening of the channels or yet another modification of the rods (the tube within the tube).

3.2.1 Why Were There No Operators on the Commission?

To appreciate the situation in the time before the accident was dominated by a drive for fuel economy. Many Soviet readers will no doubt remember the slogan of the Ministry of Energy of the USSR at that time: "Let us save 1 or 2% of our fossil fuel!", which was, unfortunately, taken up by the Atomic Energy Authority, too.

In this spirit, it was then that "under strict scientific supervision" and with the "kind permission" of the chief science officer, the steam effect at the RBMK nuclear stations reached a value of up to 4–6β. The attainment of such a high β coefficient was favored by a more complete burning of the nuclear fuel than desirable for safe reactor operation, by the removal from the core of all the supplementary moderators, and by a reduced number of chargings with fresh fuel. With such a large and fast-developing positive steam-void effect the nuclear power station staff found it naturally quite difficult to manage the huge, powerful reactor.

Such a situation cannot be attributed to a failure of the operators. Probably, that is one reason why no actual reactor operators were appointed to the commission which investigated the cause of the Chernobyl disaster. In contrast, there were plenty of the people who drew up the SRNPS and GSG regulations and then designed the RBMK reactor.

Indeed, why were there so many other specialists, but no operators on the commission investigating the cause of the accident? One can only speculate that this was done deliberately, because otherwise the recognition of the non-viability of the RBMK reactor would have been unavoidable. The gross infringements of the SRNPS and the GSG would also have been recognized. And, once these things had been recognized, it would have been necessary to shut down the remaining 14 operating reactors and carry out urgent remedial work to eliminate the faulty design features.

This is the stage at which economics again took precedence over safety. Just think, 13 of these reactors produce 1 million kW each and the other one alone produces 1.5 million – shutting them alldown at once would really hav created an energy crisis!

He was indeed right who said: "Politics is the expression of economics in concentrated form." The government itself was forced to help in the search for scapegoats among the reactor staff.

3.2.2 Who Should be the Scapegoat?

At last some use was found for the regulations, for SRNPS §5.19 states that when the reactor is operating, responsibility for the safety of the nuclear power station rests on the management, the reactor hall supervisor, and the shift supervisor.

It was precisely this clause which identified the six scapegoats. Therefore, the investigation into the causes of the accident was seemingly directed towards collecting the largest possible number of infringements committed by the staff, behind which to conceal the numerous, glaring design flaws. Meanwhile, at the remaining RBMK stations, an emergency program of extraordinary measures was started, aimed at the elimination, as far as possible, of these faults.

Here is a brief list of these measures (including those already carried out until the beginning of 1991):

1) The reduction of the steam coefficient to a value of $\beta \sim 1$. A further reduction of this positive effect, to a value close to zero, is planned.
   [Fulfilling the requirements set out in SRNPS §3.2.2 and of GSG §2.2.2.]

2) Installation of a fast-acting accident prevention system based on the standard group of emergency rods.
   [Meeting the requirements of SRNPS §§3.3.21, 3.3.27.]

3) Elimination of the positive reactivity surge which accompanied insertion into the core of the working group of control rods. This was accomplished by replacing these rods by others with a new design and by incorporating the safety rods to be inserted into the reactor from below into the accident prevention system. Their speed of operation was increased.
   [Fulfilling the requirements of SRNPS §§3.3.21, 3.3.27, 3.3.28.]

4) Installation of individual display and recording instruments which provide a constant check on the operational surplus reactivity. Also, installation of a warning and emergency alarm. There are plans to install a protective blocking mechanism with regard to the minimum-permissible OSR.
   [Fulfilling the requirements of SRNPS §3.1.8.]

5) Development of a proposed back-up protection system (BUPS) which would be different from and independent of the existing system.
   [Meeting the requirements of SRNPS §§3.3.3, 3.3.27 and GSG §§2.3.1, 2.3.2.]

6) Development of a system for diagnosis of the state of the metal of the channels and of the basic equipment of the first circuit (MFCS).[4]
   [Meeting the requirements of GSG §§4.2.1, 4.2.3.]

7) Development of measures aimed at ensuring seismic stability of nuclear power station's equipment and systems. (This question was never consid-

---

[4] "Multiple forced circulation system" (MFCS)

ered before in relation to any of the operating RBMK stations.)
[Meeting requirements of GSG §§2.1.5, 2.1.6.]

8) Release of information on the technical basis for the safety of reactor installations (TBSRI) and on nuclear power stations (TBSNPS).
[Meeting requirements of GSG §2.1.14 and SRNPS §3.1.6.]

9) Release of revised technical regulations.
[Taking into account the requirements of GSG §3.1.2 and SRNPS §4.2.3].

10) Development and introduction of measures to ensure the safety of the reactor in the event of a simultaneous rupture of the majority of the fuel channels.
[Meeting requirements of GSG §2.2.1, 2.2.3.]

11) Modification of the reactor's Emergency Cooling System (ECS).
[In accordance with the requirement of GSG §2.7.2.]

12) Development and installation of back-up emergency alarms that warn when dangerous values of certain parameters are reached (operational surplus reactivity, rate of pressure drop in the MFCS, excessive distortion of the energy output field, rupture of the RGK [5] and so on).
[Meeting requirement SRNPS §3.3.21.]

13) Modernization of the existing parameter recording program (DREG [6]) which, in case of an accident, makes it possible to ascertain the ways in which it arose and developed, and how the staff acted.
[Meeting requirement GSG §2.5.2.]

As can be seen, these measures are nearly all aimed at eliminating discrepancies between the requirements of the SRNPS and GSG regulations and the actual design features of the RBMK.

The above is by no means an exhaustive list of such discrepancies.

In May 1988, specialists from a number of organizations compiled a list of points where the RBMK failed to meet the requirements laid down in the regulatory documents, taking into account progress made with the measures given above. Their list contains about 30 deviations from the SRNPS and GSG requirements, nine of which, as is pointed out by former USSR atomic energy minister N. F. Lukonin, are technically impossible to eliminate. In the notes attached to this document we find these words:

"The present list is liable to be supplemented according to the established procedure after the publication of the TBSRI and the TBSNPS and in the event of the discovery of further deviations in the course of the operating process."

In other words, some of the shortcomings were eliminated hurriedly after the accident; a significant number (approx. 30) still exist, and certain faults (approx. 9), mainly affecting the first six RBMK blocks at the Leningrad, Kursk, and Chernobyl power stations, cannot be corrected at all.

---

[5] RGK is the Russian abbreviation for "distributing group collector".
[6] DREG is the Russian abbreviation for "diagnostic registration of parameters".

3.2.3 How Could a Reactor with So Many Defects Be Built and Put into Operation?

Firstly, no-one analyzed the RBMK plans at the design stage (that is, there was no independent, external scrutiny).

Secondly, the designers themselves did carry out an analysis, but on a very superficial level (because of the poor experimental facilities, the chronic backwardness of the available computer technology, etc.).

Thirdly, thanks to the monopoly that exists in Soviet nuclear science, the RBMK reactors, unlike airplanes, automobiles, etc., were not subjected to any serious tests or trials of their durability. That is why 16 reactors were brought on line without even a Technical Basis of Safety of Reactor Installation (TBSRI) or a TBS of Nuclear Power Stations (TBSNPS) certificate.

However, with these obligatory parts of the project missing, it is illegal to not only operate a nuclear power station, but even to build it (GSG §§1.2.3, 2.1.14). It was only in 1988 that the chief designer made an attempt to officially certify the safety of the second- and third-generation RBMK stations. This attempt is interesting in that the TBSRI certificate also includes those measures aimed at eliminating deficiencies, *which in fact have not yet been carried out, but are only planned*. In this case, the TBSRI certificate itself does not meet the standard requirements.

For the six first-generation RBMK reactors (the first blocks at the Leningrad, Kursk, and Chernobyl nuclear power stations) no such attempt at certification was made. It was rendered impossible by the extremely poor quality of the design (that is, by the nine faults mentioned above which contravene SRNPS and GSG, and which cannot be eliminated). These six reactors do not have accident isolation systems and their emergency reactor core cooling systems (ECCS) are simply parodies of such systems, having nothing but the name in common with systems which meet GSG requirements.

Thus, RBMK reactors have now been in operation for more than 15 years and only one of the 16 (the third block at the Smolensk nuclear power station, brought on line in January 1990) has a safety certificate! And yet, we are repeatedly assured that the main cause of the Chernobyl accident was the "human factor" – meaning only the mistakes made by the reactor staff.

3.2.4 "Human Error" – A Well-Defined Term

In any textbook (for example *Monitoring, Control and Protection Systems in Nuclear Power Stations* published by the Obninsk Institute of Atomic Energetics, 1987) the term "human error" is used to include mistakes made by people at all three stages of the process which should ensure nuclear safety:
- planning and design (stage 1);
- manufacture and assembly (stage 2);

- on-line operation (stage 3).

Why is it, then, that the accident investigators targeted only the last stage? Why was it that in the course of investigating (or rather selecting) the courses of the accident they forgot the first, most basic stage – the stage at which the safety level of the future nuclear power station is determined?

Actually, we have already answered these questions above.

In the same textbook it is stated:

"The practical realization of safe operation depends, to a large extent, on the safety features built into the design. It must be remembered, however, that, during manufacture and operation, safety standards usually only drop by comparison with the design standard."

3.3 The Six Alleged Infringements of the Reactor Staff

Let us look at all six of the *"most dangerous infringements of working practices"* committed by the staff of Block 4 of the Chernobyl Nuclear Power Station, which amounted to the *"freak combination of infringements"* considered to be impossible by the designers of the reactor and which prompted them to conclude that they could *"not envisage the creation of safety systems capable of averting the accident"* (from the information presented to the IAEA).

Let us analyze each of these "infringements" from several points of view:

1) Did they really happen, or were they "invented" by the experts?
2) If the infringement in question really is an infringement of the regulations, as laid down in any of the relevant documents and which the staff should have known and adhered to, then was it in fact a cause of the accident? How did it affect, or would it have affected the course of the accident?

3.3.1 The First

The first of the staff's infringements, as formulated in the official sources (reports to the IAEA and to the public is:

> *"The reduction of the OSR to a value substantially lower than that permitted."*

Experts are of the opinion that, in consequence:

> *"The reactor's emergency protection system was rendered ineffective."*

This point has already been extensively addressed, but let us summarize and elaborate a bit further.

In the regulations (see Chapters 10, 11) concerned with the duties of staff in regimes involving big reductions in reactor power (regimes ES-1, ES-2, ES-3,

ES-4, manual rundown [7]) there is absolutely no mention of the need to monitor the operational surplus reactivity among the list of basic parameters which must be monitored (§8.9.1.)

The Technical Regulations did not (and, in fact, even today do not) state what action the staff should take in the event of a computer malfunction affecting the operational surplus reactivity read-out (the standard "Prisma" program).

All this, however, does not imply that the staff paid no attention to this parameter. A limit is a limit, whatever the process by which it is fixed, and it has to be observed. But, the question arises − did the staff really go beyond the limit?

The scientists said that they did − and significantly, practically going down to half the safe level.

However, we cannot be sure as there are no printouts showing such an operational surplus reactivity - which demands the question, "Did the staff see an operational surplus reactivity (OSR) readout of less than 15 before the start of the experiment?"

We only know (from the evidence of witnesses) that 30 minutes before the start of the experiment the OSR value stood at 18–19 rods. The cited OSR value of 6–8 rods was obtained by the investigating scientists as a result of calculations based on the original data as recorded on magnetic tape by the standard computer.

If we now assume that the Chernobyl staff "blinked and missed" changes in this parameter (although admittedly the design faults strongly facilitated such a lapse), then how are we to understand the recommendation of the chief designer which we considered in the first part of this chapter?

According to this recommendation, in order to maintain the effectiveness of the accident protection system on operational surplus reactivity, only 2.5–5 rods are sufficient.

Thirdly, if again we assume that the operational surplus reactivity limit was exceeded, how can we accept a situation in which the violation of a single parameter makes it impossible for both active groups of emergency rods to stop the reactor − not just to fail, but to actually cause its runaway? In other words, if the OSR value was indeed violated by the staff, then this is primarily a direct consequence of the deficiencies of the design and of the poor quality of the operational documentation.

---

[7] ES is the abbreviation used for Emergency System; its Russian abbreviation is AZ.

3.3.2 The Second

The official formulation of the infringement is

> *"A power output drop to a level lower than that envisaged in the experiment plan."*

The consequences of this infringement, in the opinion of the scientists and experts:

> *"The reactor became difficult to control."*

Indeed, the staff did depart from the experiment plan, which decreed that the power output during the experiment should be 700–1000 MW. However, not in a single document dated before April 26, 1986, is there any mention of a minimal power output limit.

Moreover, all the RBMKs were equipped with the blocking mechanism included in the original design (regimes ES-3, ES-4) which should automatically shut down the reactor at power output below 640 MW (20% of nominal).

Besides, the Technical Regulations (§11.4) require the staff in such situations (disconnection of the reactor from the grid) to bring the thermal power output of the reactor into line with the electrical load of the block's own needs – that is, to reduce under manual control the reactor's power down to a level of 200–300 MW (the electrical load of the block's own needs is estimated at 70–80 MW, the efficiency of the block being of the order of 30%).

Similar reductions of power by the operators are governed also by other sections of the Technical Regulations (the action of lowering or raising frequency of the voltage in the grid). The regulations governing experiments and tests of the emergency safety systems (§5.6.1.1) as approved by the chief science officer and the chief designer instructed that testing of the main safety valves should take place with reactor power at less than 700 MW and with turbine generators switched off.

The "infringement" in question, then, is a deviation from the staff's own experimental program but not from the requirements of the regulations which ensure the safe operation of the reactor.

Before the accident, there was no mention in any operating document of such a limit (i.e., 700 MW) on safe operating conditions. Only after the accident were the blocking mechanisms of the original design (ES-3, ES-4) hurriedly disconnected and a minimum power limit of 700 MW established, down to which a reduction of power is permitted, but below which the reactor must be tripped immediately.

It was not scientific calculations, but bitter experience which showed that the RBMK reactor, apart from its defective accident protection also suffers from dangerous power output levels, at which it becomes "difficult to manage".

The scientific supervisors (Institute of Atomic Energy) in their report acknowledge their ignorance of the dangers of operating the reactor at low power output levels:

"In this sense the Chernobyl accident has been very instructive. It turned out that the severity and proportions of an accident at low output levels can be substantially greater than at high output levels. A purposeful study of possible accidents, even if some of the possibilities seemed highly improbable, might have guided the researchers towards a realization of the special danger inherent in such power output regimes . . . ".

What exactly, is this special danger inherent in low output levels? And what part did it play in the origin and development of the accident?

The fact is that behind this vague statement that "the reactor becomes difficult to manage at low output levels" lies yet another attempt to conceal a glaring infringement of the SRNPS §3.2.2. and of the GSG §2.2.2 by the original design. The rules state that the power coefficient of reactivity must not be positive, whatever the state of the reactor – and if it is positive, then the designers must ensure and give firm proof of the safety of the reactor in stable, transitional, and emergency regimes.

This has never been done to this day.

On April 26, 1986, the positive steam-void effect at lower power output levels was greater than the negative temperature effect of the fuel (Doppler effect), the sum of the two creating the rapid effect of the reactivity on the power output. Thus, basically, the special danger inherent in such a regime lies in the positive feedback between reactivity and power output. In other words, the RBMK reactor in this state had self-accelerating properties.

3.3.3 The Third

The official formulation of the infringement is:

> *"The connection to the reactor of all main circulation pumps (MCP) and the exceeding of the limits for individual MCP as laid down in the regulations."*

Scientific and expert opinion as to the consequences:

> *"The temperature of the coolant in the multiple forced circulation system (MFCS) approached saturation temperature."*

The connection of all the MCP to the reactor, regardless of power output level, was never forbidden by any of the documents, including the technical regulations manual, before April 26, 1986. The limit on the number of pumps permitted to be connected to the reactor at power levels below 700 MW was laid down only after the accident. There is, therefore, no question of any infringement on the part of staff in this regard.

At low power output levels (below 700 MW) when the input of water is less than 500 tons per hour, there was, and still is, a limit on the throughput of the main circulation pumps (MCP), i.e., not more than 7000 cubic meters per hour.

This restriction was introduced *"in order to ensure adequate supply up to cavitation"*, see the Technical Regulations.

Yes, in fact, certain of the pumps were, at the time of the experiment, working at a rate of about 8000 cubic meters per hour (in contravention of point 11.1.6 of the Technical Regulations). But this, as we can see, exceeds the limit for such a regime by no more than 20%. The important question then is: did cavitation of the main circulation pumps (MCP) take place on April 26, 1986?

No, it did not. This is made clear by the hard data that are available in the form of print-outs from the diagnostic recording system (DREG), which, as the Institute of Atomic Energy scientists note in their report, are the most objective. This is confirmed by the conclusion arrived at by the scientists themselves in the same report: *"The results of hot-water tests on the MCP pumps show that, in the conditions in which the Chernobyl MCP were working on April 26, 1986, cavitation of the pumps cannot take place."* Information about the conditions in which the Chernobyl main circulation pumps were working on April 26, 1986, has been supplied by the factory which manufactured the pumps.

This is one, out of 13 possible causes of the acceleration of the reactor that were considered by the Institute of Atomic Energy scientists. They themselves rejected it as a realistic possibility. Therefore, this infringement on the part of the staff was obviously not the cause of the accident.

The danger inherent in the situation which arose on April 26, 1986, did not stem from the connection of all main circulation pumps, or from the fact that pumping limits for individual main circulation pumps were exceeded, but rather from the large overall expenditure of coolant through the core at a low power output of the reactor. When there is such a large flow of coolant and a low power output (neither of which, unfortunately, was banned by the regulations) conditions are created for the maximum steam-void effect to arise, having in this case a high value of $4 - 6 \ \beta$.

3.3.4 The Fourth

The official formulation of the infringement is:

> *"The blocking of the reactor's protection system when the indicators showed that both TGs (turbine generators) had stopped."*

According to the experts, the consequence of this was:

> *"The loss of the reactor's automatic trip facility."*

Let us first clarify the formulation of the infringement as cited above. It must be taken to mean the intentional deactivation of the reactor's automatic stop circuit when the one working turbine generator had been switched off.

Yes, it is true that the staff intentionally deactivated the reactor's automatic stop circuit, assuming the right to take manual control by pressing the emergency button. Did they in fact have this right? There was no mention of this matter in the experimental program.

The Technical Regulations gave (in chapter 17) the following instruction concerning the staff's operation of protection and alarm systems:

"As regards the connection and disconnection of protection and blocking mechanisms and of alarms, the staff must act in accordance with the regulations governing the positions of the switches and cover plates of the protection and blocking mechanisms."

But the regulation governing the use of switches (§1) states:

"The reactor's protection switches must be in the OFF position while the two turbine generators are being stopped and when the electrical load has dropped to 100 MW or less."

The power output of turbogenerator No.8 was 50–60 MW before the start of the experiment (TG-7 had been stopped earlier). The staff, therefore, did not break any of the regulations.

Did the fact that the protection system was switched off have any bearing on the start or the development of the accident?

No, it did not. It did not, since the conditions for the runaway of the reactor already existed before the beginning of the experiment. The reactor could have run away 36 seconds before the staff pressed the emergency button, at the moment when the "both-generators-stopped" protection system was activated.

Concluding the question arises as to the difference it makes whether the reactor exploded because the emergency button was pressed by the operator or because the automatic protection system went into action?

3.3.5 The Fifth

The official formulation of the infringement is:

> *"The disconnection of the safety mechanisms relating to water level and steam pressure in the steam drums."*

The opinion of the experts as to the consequences of this action was:

> *"The reactor's protection as regards thermal parameters was completely de-activated."*

It must be said immediately that the official formulation of this "infringement" and especially of its consequences looks like pure disinformation.

What safety systems relating to the parameters in question were actually disconnected? There is a short answer: none. They all continued to operate. The staff did not even touch the safety controls relating to maximum water level (+250 mm) and steam pressure (75 atm) in the separator drums. These mechanisms remained in operation, at the indicated settings. Only the setting of the protection mechanism relating to minimum water level in the separator drums was not changed by the staff from −1100 mm to −600-mm (in accordance with the requirement of the *"Regulations governing the positions of switches and cover plates ..."* the change of setting should be made when the power output falls below 60% of nominal).

Can it be said that the protection system was disconnected completely?

Of course not, as any specialist will confirm. The separator drums' minimum water level protection system remained in operation, but was set roughly 500 mm lower.

Did this infringement have any bearing on the start or development of the accident? No, it did not. It did not, since the water level in the separator drums directly before and during the experiment hardly changed at all and was above the −600 mm level.

Let us consider the protection mechanism relating to the reduction of steam pressure in the separator drums. It must be noted that it works in steps: when steam pressure drops to 55 atm, one of the two generators (which one is not important) is switched off; on a further drop to 50 atm the remaining generator is also switched off. The *"Regulations governing the positions of switches and cover plates ..."* give the staff the right to decide which generator to disconnect, and at which separator drum steam pressure setting it should be disconnected. Exercising the right given to them by the regulations, the staff changed the setting from 55 to 50 atm for turbogenerator TG-8.

The separator drum steam pressure directly before and during the experiment hardly changed at all, while after the emergency button was pressed, it rose sharply. Moreover, if the experiment had been carried out on turbogenerator TG-7 instead of TG-8, then the above-mentioned change of setting would not have been necessary since the setting for TG-7 was already 50 atm. The reactor's protection systems, therefore, remained in operation at normal settings.

In summary, not one of the considered separator drum steam pressure and water level safety systems was disconnected. All remained in operation. The only infringement involved here was the altering of the setting of the separator drum minimum water level protection system.

3.3.6 The Sixth

The official formulation of the infringement is:

> *"The disconnection of the Maximum Design Based Accident Protection System."* [8]

In the opinion of the experts the result was:

> *"Loss of the ability to reduce the scale of the accident."*

The Maximum Design Based Accident (DBA) is an accident resulting from the most serious of all initial events which, in the case of the RBMK, is

---

[8] The disconnection of the emergency reactor core cooling systems (ECCS).

(according to the GSG) taken to be the instantaneous transverse rupture of the main circulation pump pressure collector with the maximal internal diameter (900 mm). If this were to happen, the result, whatever the working rate of the pumps, would be an immediate dead stop of the flow of coolant to the fuel channels, which would be empty within 2 seconds.

The emergency core cooling system (ECCS) plays no part in the energy production process. It remains in a constant state of readiness and, if there is any sign of a rupture of a large-diameter (over 300 mm) MFCS pipe, it is supposed to prevent the emptying of the fuel channels.

Why? Because, among other deficiencies, the ECCS has another one which is substantial: it admits cold water (of not more than 40°C) into the MFCS pipes and fuel channels, which are heated to about 300°C. This could lead to a further rupture of the pipes because of the inevitable thermal shock and the excessively rapid cooling of the metal (the maximum permissible emergency cooling rate is less than 30°C per hour).

It is exactly this fact which caused, and still causes, apprehension among the staff concerning what would happen if the emergency reactor core cooling systems were put into action by mistake. After all, it is one thing if rupture of a large-diameter MFCS pipe really has taken place and there is an urgent need to prevent a meltdown of the fuel by means of an immediate delivery of the necessary quantity of water (even if the water is cold), and quite another matter if this cold water from the emergency core cooling system is delivered into the hot MFCS pipes when there is no need for it – with the possible consequences already indicated.

For this very reason, a full inspection of the emergency reactor core cooling system (ECCS) is carried out, according to operating instructions, only when the reactor has been stopped and allowed to cool. However, a partial check (of one section) can be done while the reactor is working, as long as no cold water is admitted. In order to inspect a section of the ECCS on a working reactor without admitting cold water, it is necessary to remove that section from its state of readiness for the duration of the test.

For how long is this permitted?

Before April 26, 1986, the permitted period was not laid down in any of the instructions. The reason for this was something already familiar to us – the lack of any TBSRI or TBSNPS technical documents for this type of reactor, the lack of any regulations governing the services and inspection of safety systems.

According to the *"Rules governing technical operation ... "* (§29.23) the permissible period in such cases, as well as in cases where malfunctions of the emergency reactor core cooling systems (ECCS) are discovered, should be determined by the station's Chief Engineer. The Technical Regulations gave no specific ruling on this matter and, in fact, give no such ruling even now. There was only one requirement – that the ECCS should be brought into a state of readiness if the temperature in the MFCS exceeded 100°C. In the SRNPS, however, §5.4. requires that the emergency cooling system should be in working condition when the reactor is started up – and does not specify clearly whether this refers to the ventilation and cooling system (VCS) or the emergency reactor core cooling systems.

Let us remember that, during the experiment, the ventilation and cooling system, with its emergency cooling capability, remained in operation. For the purposes of the experiment it was sufficient to disconnect only one (the third) of the three emergency reactor core cooling systems channels. According to the plan, its fast-acting section should receive its electricity and working medium (feed water) by the generator as it spins to a halt.

However, because of fears of a false activation of all the ERC channels, and given the short duration of the experiment (about 4 hours according to the plan), and taking into consideration the shutdown of the power plant for repairs which was scheduled to follow the experiment, the chief engineer considered it permissible to disconnect the emergency reactor core cooling systems (ECCS) completely. In fact, this was an infringement of the above-quoted §5.4 of SRNPS and the requirements of the Technical Regulations.

The following questions arise: To what extent was the reactor rendered less safe by this action? What part did the disconnection of the ECCS play in the start and development of the accident? According to the hard data, the pressure in the MFCS during the experiment, right up to the pressing of the emergency button, hardly fluctuated – but after the button was pressed it rose sharply and was higher than that in the emergency reactor core cooling system tanks.

If, instead of the whole emergency reactor core cooling system, only the third channel involved in the test had been disconnected, then the scale of the accident would not have been reduced. Firstly, the minimum reaction time of the emergency reactor core cooling system (time taken for the water to reach the core) is 4 to 5 seconds. This is the same as the time it took the reactor to run away, with the accompanying sharp rise of pressure in the core. Secondly, before the reactor exploded there were no signs of a rupture in the MFCS pipes and therefore – and has already been proved – no automatic activation signal was transmitted to the emergency reactor core cooling system (ECCS). After the explosion the ECCS was superfluous – all the more so because it had been destroyed.

3.4 Conclusions

Let us sum up the "infringements" committed by the Chernobyl staff on April 26, 1986:

- the operation of several (two out of eight) circulation pumps with a slight (no more than 20%) overload (an infringement of the Technical Regulations);
- the lowering of the setting on the separator drum minimum water level protection mechanism by 500 mm (an infringement of the *"Regulations governing position of switches and cover plates of technical protection systems"*)
- the disconnection of all the channnels of the ECCS, instead of just the one involved in the experiment (an infringement of the Technical Regulations and the SRNPS).

Not one of these infringements, or even all of them in combination, caused the accident or affected the course which it took, or the scale of its consequences.

The cause of the accident lay in the grievous deficiencies of the reactor's design, the most basic of which are:

- the dangerous physical characteristics of the core (the large positive steam-void effect of reactivity and the positive power effect at low output levels);
- the defect inherent in the reactor's emergency protection system (the occurrence of a positive surge of reactivity when the emergency control rods are inserted from their extreme upper position);
- the extremely slow operation of the emergency protection systems, and the lack of any fast-acting protection system;
- the poor quality of the design documentation and, hence, also of the operating documents.

Since the accident, even the thought of using a turbogenerator as it spins down gives rise to gravest concern. But the designers intended this as a way of ensuring that the fast-acting section of the third channel of the emergency reactor core cooling system could continue to work if an accident in the grid, causing power failure at the nuclear station, ever coincided with a Maximum Design Based Accident.

Unless this "spin-down" regime can be used in the cases mentioned, taking into account the requirements of SRNPS-82 (the principle of unit failure), the RBMK reactors now at work and their three-channel emergency reactor core cooling system cannot operate legally at a power output level of more than 50% of nominal. Nevertheless, they continue to do so.

The six original RBMK reactors at Leningrad, Kursk, and Chernobyl are well into their second decade of working practically without any emergency system (ECCS) and none of the scientists seems interested in analyzing the scale of a possible future accident.

Incidentally, as previously, the new SRNPS-88 gives sole resposibility for safety to the operating staff of a nuclear station while the chief designer and the chief science officer again remain on the sidelines.

At this stage, the reader might question, "It is easy to fool the public – but what about the IAEA conference of August 1986, attended by hundreds of foreign experts? How could they have agreed with the conclusions presented by the Soviet delegation headed by Academician V. A. Legasov?"

This question can be answered by an excerpt from the report of the International Consultative Group on Nuclear Safety (an IAEA group) written in September 1986 at the request of the director of the IAEA on the basis of information supplied by Soviet specialists:

"Most of this report is taken from the well-prepared information presented at the conference by the Soviet experts ... In normal operating conditions the power coefficient of the RBMK at full power is negative and becomes positive when power output is reduced to a level below about 20% of the full power. The operation of the reactor at an output level below 700 MW is restricted by rules of operation, in view of the problems involved in maintaining the thermohydraulic parameters in the normal working range ... ".

In order to ensure the required distribution of the energy output and the effective use of negative reactivity in emergency conditions, the rules prescribe that *"no fewer than 30 effective rods must remain inserted into the reactor's core ... "*.

The occurrence of a positive reactivity surge that may be caused by the system of control and emergency rods is not even mentioned in the report. However, the occurrence of a positive power coefficient of reactivity at low power output levels is acknowledged.

Let us repeat once more that before the accident not a single document forbade operation of the reactor at a power output level of less than 700 MW and with an operational surplus reactivity of less than 30 rods; there was no mention in any document of the occurrence of a positive power coefficient of reactivity.

These restrictions were introduced only AFTER the accident. As we can see, the "well-prepared information" supplied by the Soviet experts can be considered as misinformation. If we add to this "information" the (to our knowledge, wrong) statement that the reactor staff disconnected all the protection systems relating to water level and steam pressure in the separator drums, leaving the reactor totally without protection regarding its thermal parameters, then we cannot help getting the feeling that the investigation of the accident was directed along the wrong path. If that really was the case, was the aim to conceal the true causes?

Many questions remain open, including that of who are the culprits and who are the victims. One thing is clear, as long as matters of nuclear safety are treated in the way described above, it is impossible to rule out the repetition of such accidents. Only a truthful, public exposure of the causes of the Chernobyl tragedy (and of others) will save us from wrong regulatory decisions, and secure effective public control in matters of safety. We all know that each of our actions is associated with a certain risk, however, at each step we should be aware of the risk we are accepting as being reasonable or inevitable.

4. The Zone

"In Kiev, when we deplaned, the first thing that we saw was the cavalcade of black government automobiles and the anxious crowd of Ukrainian officials. They had no exact information available, but they said that the situation was bad. We piled into the cars and drove to the power plant.

"I must say that I had no idea at the time that we were driving towards an event of global proportions, an event which it seems now, will go down in history like the eruptions of volcanoes, the destruction of Pompei, or something like that."

Academician Valerii Legasov
"Moi dolg rasskazat' ob etom" ["My Duty to Tell About This"] Phantom (Molodaya Gvardiya, Moscow 1989) p. 8

The Ministry of Mechanical Engineering (MinSredMash), the ministry which designed the dangerous RBMK nuclear power blocks, and the entire Soviet Union, were not prepared for a catastrophe like Chernobyl. (Are others better prepared?) There was no clear plan of emergency action and none of the necessary technical instruments and robots. There was no professionally trained Emergency Team that could quickly localize such large-scale radiation pollution and effectively deal with it.

Special groups, which by design are intended to enter ground zero in case of a nuclear explosion, were not provided with protective clothing and had neither the tools nor the methods to ensure that the radioactive nuclides would not spread.

Everything was organized as events progressed: the work of the Goverment Commission itself, and that of various committees and of more than 40 ministries and institutions.

Preceding page, the area within 30 km around Chernobyl nuclear power station is now closed to the public. The sign carried by the two men in the photo reads "PROHIBITED ZONE".

The country unleashed all its resources to deal with the disaster. Billions of roubles and enormous human effort were expended. Trucks bringing equipment (including that ordered and even some unasked for) stretched for hundreds of kilometers from Chernobyl. There was simply not enough time to unload them.

The most anguishing aspect, however, was the number of people – hundreds of thousands – who were torn away from their families and their work. People were called up for emergency duty, often at night, with no regard for their health or obligations, then dressed in military uniforms, and without any preparation, training, or protection they were thrown into the hell of the Special Zone (the 30-km Zone around the accident site). With shovels and with their hands they had to gather the radioactive graphite which the destroyed reactor had thrown out onto the roof of the neighboring Block 3.

By the end of May 1986, the regular (drafted) troops in the Zone were starting to be replaced by reservists; more than 650 000 people passed through the Zone. Without appropriate dosimetry and medical monitoring they began to build the "Sarcophagus" (i.e., a containment building around the wrecked reactor).

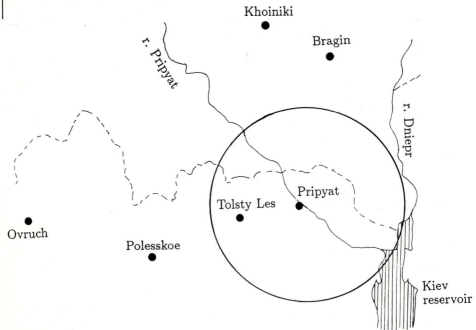

Fig. 4.1. The Zone

4.1 Organizational Structure and Radiation Safety

The biggest difference between the Chernobyl Cleanup and projects of similar scale is that in the latter the simple problems were usually not planned in advance, but dealt with as they arose. Under the conditions that existed at Chernobyl, however, a mistake in judgement in even the smallest job could mean significant problems at a later stage. In this chapter the organization of the Cleanup of the Zone and the problems encountered are discussed.

4.1.1 Organization of the Rectification Project

A special Government Commission headed the coordination of the diverse ministries, primarily the Ministry of Defense, MinSredMash, and the Ministry of Energy, and other departments of the "Rectification of the Consequences of the Accident" (Russian: "Likvidatsiya posledstvii avarii" – LPA) project. Each of the engaged ministries had as its primary administrative body an "operational group" (OG) whose directors were members of the Government Commission. Thus, it was necessary to coordinate the activities of various groups and to provide for their harmonious interaction.

The territory affected by the accident was divided into several subregions, one of which was the inner 30-km radius circle around the plant, the so-called Special Zone. This area was managed by its own operational group (OGSZ), but was a subunit of the Operational Group of the Ministry of Defense (OGMD).

Another task of the Commission was to acquire the finest specialists (especially in nuclear power station matters) to identify the best course of action. The two most important complicating factors in finding appropriate personnel were that most of those involved and hired had no knowledge of appropriate methods for working in ionizing radiation fields and no familiarity with radiation safety practices.

The operational group of the Special Zone managed the project with flexible time planning, but with an eye to the original time schedule. In undertakings similar to this one, in terms of its scale and of the constantly changing parameters, such a scheme is certainly applicable, but it requires well-trained and -prepared personnel with original ideas, high analytical skills and ability to make immediate decisions under pressure. The members of the operational group certainly did have most of these qualities, but had no special training or preparation for such a situation.

4.1.2 Radiation Safety

At the end of 1986, the production union "Kombinat" was created within the Ministry of Energy by decision of the Government Commission. It turned out that this decision was not a wise one. Among the reasons for this conclusion was the nature of the problem itself. This union was supposed to be extradepartmental, and all organizations and undertakings of all ministries and departments within the rectification work were to report to it. But a number of complications in the enforcement of radiation safety procedures arose as a result of the creation of Kombinat.

One of the subdivisions of Kombinat was the Dosimetry Monitor Management (DMM, below we will refer to it as Dosimetry), which was supposed to be the main body in the Zone charged with dosimetric monitoring and radiation safety. However, yet another mistake – this time by the Ministry of Atomic Energy – was made in the decision to separate the radiation safety program of the Chernobyl nuclear power station (ChNPS) from the Dosimetry. The Dosimetry (DMM) was also subdivided into the following divisions: the department of individual dosimetric monitoring, the department of external dosimetry, the department of radiation safety, the department of data analysis and prognosis, the department of dosimetric instrument and apparatus repair, an engineering department, a bookkeeping department, and a personnel department.

In view of the immense problems facing the Dosimetry, hiring was intense, and not selective. This did not seem to concern the Dosimetry; its greater problem was that it was expected to function at full capacity. In a short time (until August 1987), the new staff was trained and allowed to begin independent work, basic standardizing documentation was prepared, and the recommendations therein were put into practice by decisions of the operational group of the Government Commission.

In addition to the Dosimetry, the Government Commission decided to create a radiation safety inspection board, which was under the direct command of the Ministry of Atomic Energy. The main job of the inspection board was to coordinate the radiation safety programs of the various departments. Placing the board directly under the Ministry of Atomic Energy was also a mistake.

Of course, the inspection board did contribute a great deal to the progress of work in the Zone, but in practice it could only issue declarations and did not have real power.

This intricate organizational structure in radiation safety matters led to a great many problems being left unsolved. The main tasks of the Dosimetry were

- monitoring of radiation levels in areas of permanent habitation in the workers' communities (all workers except for Ministry of Defense and MinSredMash personnel);
- preventing the spread of radiation contamination beyond the Zone;

- ensuring sanitary conditions in communal eating areas, in the living quarters of the worker communities (all workers except for Ministry of Defence and MinSredMash);
- organizing the work projects in radiationally dangerous areas in the Zone with the exception of those projects carried out by the Ministry of Defense and Ministry of Engineering (MinSredMash);
- returning of building equipment and land to use after decontamination
- monitoring waste-storage facilities and developing measures to prevent leakage of radionuclides into surface and ground waters;
- monitoring radiation exposure of all personnel except for those of the ChNPS, the Ministry of Defense, and the MinSredMash;
- developing and maintaining measures to limit the exposure of personnel;
- developing plans to localize radiation contamination of structures in the surrounding environment and organizing their implementation;
- developing measures to increase the effectiveness of the decontamination projects and ensuring their efficient execution;
- bringing the NPS radiation danger monitoring system into operation and modernizing it.

4.1.3 The Chief Engineer of Dosimetric Monitoring

Dmitrii Vasilchenko, Chief Engineer of the Dosimetric Monitoring Management (DMM in here referred to as Dosimetry) of Kombinat explained to me the organization of the radiation safety service from 1986 to the present. He worked in the Zone since 1986 on the decontamination of Blocks 1, 2, and 3, and was with us on the roof of Block 3 during one of the most dangerous phases of the removal of highly radioactive debris. In a word, he is an expert in radiation safety, particularly on the huge scale of what was required at Chernobyl. His statements are as follows.

"The main problems involved in maintaining radiation safety standards were:

1) Insufficient numbers of dosimetry apparatus and laboratory equipment.
2) Insufficient numbers of personal dosimeters with broad ranges of sensitivity.
3) Lack of an automated system for control of personal dosimeters (not just at Chernobyl, but in the entire USSR), which could also be used as the basis for a data bank on individual radiation exposures for later analysis.
4) A poorly staffed permanent system of radiation monitoring and insufficient numbers of data collectors ready to deal with such an accident. The permanent radiation monitoring staff at Chernobyl was unable to provide reliable information to the operators.
5) Lack of state-of-the-art devices to measure internal radiation levels, lack of means of calculating the internal radiation doses.

6) Inadequate radiation safety training and an almost complete lack of basic instruction in radiation safety among the residents of Pripyat. In my opinion, one result of this is that significantly more numbers of people were irradiated than is justifiable.

7) Too few trained dosimetrists, which made intelligent task organization after the accident practically impossible.

8) Inadequate attention paid to worker protection and safety by the nuclear power station administration (the director and the chief designer). Although this may at first appear unimportant, in my opinion this had a significant negative influence on the course of the accident.

Staff discipline, the strict following of all instructions, rules, and regulations, is very much dependent on the level of support and leadership demonstrated by the department of worker safety and protection, and especially by the nuclear power station director. My long experience in these matters has convinced me that the department of worker safety must be headed by the director and no-one else.

9) The lack of an automated system of radiation safety at all nuclear facilities, and Chernobyl in particular.

My analysis of the practices of the radiation safety workers, conducted these last 4 years, has shown that these nine points prevented the workers from acting professionally, although in isolated cases they would have been able to and should have proceeded with more confidence.

Another reason for the large number of people exposed to high radiation levels was our so-called 'patriotism' effect. The combination of this effect, the ignorance of the nuclear power station personnel and persons brought in to work on the rectification project, as well as a practical lack of knowledge of radiation safety rules, caused the number of radiation victims to increase further.

The evacuation of Pripyat and the Chernobyl nuclear power station personnel meant that only a small group of trained people was left at the station to man the three stopped reactor blocks. The worker protection and safety staff were practically all removed from the station. In this situation the level of monitoring to ensure safe working conditions was rather rudimentary. The Government Commission divided the task of radiation monitoring among the various ministries, in principle, each ministry was responsible for its own people. This seemed to be the wisest decision at the time.

Later, however, the drawbacks of this conclusion were realized, namely, that because of the lack of coordination among the ministries, invaluable information on exposure levels of a large number of workers has been lost. This is one of the primary causes of anger among ex-rectifiers.

In mid-May 1986, a radiation safety service was founded at the Chernobyl nuclear power station. This service was put together hastily, in a disorganized fashion, with people being brought in from other power stations. Since the tours of duty of these people were 1 to 15 months long, and because they were mostly from WWR facilities and unfamiliar with the setup of the Chernobyl nuclear power station, this department was not very effective, and not noticeable in relation to all the other work being done.

I should note, however, that the directing of this department by the Vice Chief Engineer of the rectification work was quite competent. He arranged the recruitment of highly qualified dosimetry specialists, who remained on the site until October 1986. Having these specialists, in my opinion, allowed us to keep the exposure levels of rectification work personnel to a minimum. This department performed a large portion of the ground work with the help of military personnel and a small group of civilians.

The greatest difficulty with organizing an effective program of radiation safety was the job of monitoring the doses of such a large number of people. I mentioned that the Ministry of Engineering (MinSredMash) and the Ministry of Defense were responsible for monitoring the radiation exposure of their own staffs. The Ministry of Engineering had a rather good, quite professional setup for measuring individual doses. The Ministry of Defense, however, had basically only a collective radiation exposure monitor, and used calculations of the gamma radiation power. No exposure data on the military personnel for 1986–1987 are available; they were lost.

The exposure levels of the personnel of all other ministries were monitored by the group of the Chernobyl nuclear power station. But this program was organized to acceptable standards only towards November/December 1986. This was because of the complexity of the task: a large number of different organizations and institutions, a large number of individuals to monitor, no regular shift hours in the Zone (sometimes around-the-clock, or extended work for a month or more, departure, and then return to the Zone, etc.), poor knowledge of radiation safety practices and unprofessionalism, low levels of discipline and responsibility, as a result of which the personnel of the Ministry of Coal Industries, the Ministry of Energy, and other departments, in most cases, worked without any individual dosimetric control in high radiation fields, which to this day is the cause of much discontent among those who worked at Chernobyl in 1986.

The solution of this problem would have been impossible without state-of-the-art computing methods, on the basis of which a unique system of individual dosimetric monitoring was developed and put into operation, and which is still being used.

I should note that during 1986, because of the harsh radiation conditions, most of the radiation monitoring was based on measurements of gamma-radiation power. This monitoring was done by the Ministry of Defense at all controlled gates to the Zone, on all equipment leaving, using transportable

DP-5 dosimeters. No automated dosimetry measurements were made except at the 'Dityatki' gate, where in 1986, a radiation control station for vehicles was designed and constructed by the Institute of Nuclear Physics of the Academy of Sciences of the Ukranian SSR. [1]

I'd like to say a few more words about the individual dosimetric monitoring, which is the most important task of the Dosimetry. I mentioned that by decision of the Government Commission the Dosimetry was divided among the Ministry of Energy, the Ministry of Defense, and the Ministry of Engineering (MinSredMash). The main problem was with the Ministry of Defense – there are basically no questions about the activities of the other two. The problem is not an easy one, and will become worse with time as the complaints of the rectifiers who were brought in for a short period become more grievous.

It is known that by government decision certain privileges were established, among which was the one-time payment of a five-fold salary to persons receiving a dose of 5 AED (Allowed Equivalent Dose, see also Appendix A) or more. These supplements were paid out in 1986, 1987, and to this day. This significantly complicated the dosimetric monitoring. At the end of 1986 and in all of 1987 many cases of exposures of over 25 Rem were reported. Each of these cases were investigated and each of these individuals was sent to the clinic MS-1226 for further medical inspection. In the town of Chernobyl, a laboratory at the Vavilov Institute of Genetics was established to conduct cytogenetic studies of chromosomal aberrations. As a rule, large irradiation doses were NOT confirmed by these studies.

Now, about the monitoring equipment. At first glance, there seems to be enough dosimetric and physical laboratory apparatus on the market. However, except for rare exceptions, these instruments are not adequate for the conditions in the Zone. In the 4 years after the accident, no new equipment which would be adequate and convenient to use in such conditions has been produced, or even been developed (except for the DRG-01-T, which measures only a single parameter, the gamma radiation power). There are no technical means for monitoring the levels of contamination of humans by beta-emitting nuclides. As a result, at all gates to the Zone nonstandard apparatus designed by the Ukranian Academy of Sciences Institute of Nuclear Physics are used. The same is true of physical laboratory equipment. Today, the Dosimetry (DMM) uses in its routine tasks only imported devices – domestic instruments are notoriously unreliable. The technical devices for monitoring internal irradiation are not even in domestic production.

Four years of experience have shown the following:
- Right from the initial moment of the accident, serious mistakes in the organization of radiation safety, in particular, separating the tasks among so

---

[1] See also Section 4.5.2.

many departments and ministries led to incongruity and miscommunication. The large number of groups responsible for dosimetric monitoring were without any coordination, which led in many cases to contradictory data.

- Placing military personnel beyond the bounds of the Zone contributed to the further spread of radioactivity.
- The local population was not prepared for a nuclear accident at the neighboring power plant.
- The radiation safety personnel was very poorly equipped and lacked the necessary apparatus.
- Overstaffing of the rectification personnel prevented the establishment of a rigorous radiation safety training program.
- There was a severe shortage of personnel airlocks, especially in 1986.

I believe, based on the experience of these four years, that it is absolutely necessary to create a national institution, with a minimum number of employees, equipped with a wide range of apparatus, a large body of documentation, for the eventuality of a similar accident occurring anywhere in the world. This institution should have appropriate status, with the right to put new technologies into series production, to remove old equipment, to develop standardizing documentation, methodologies, and instructions. This institution should also be charged with organizing emergency procedures after an accident involving radioactive discharges."

4.2 Decontamination and Rectification

The largest task force was that of the Ministry of Defense, numbering 40 000 persons, including over 5000 officers, and 10 000 technicians. Depending on the size of the area being worked on and the speed with which a task was to be completed, the number of troops present at any given time varied between 5000 and 10 000.

Unfortunately, it was impossible for the Ministry of Defense personnel to carry out the decontamination project, especially in the plant interior, without the advice and oversight of trained nuclear power station personnel. Using the necessary equipment, both electronic and mechanical, required special training and supervision. Most of the Chernobyl personnel had been evacuated, except for a small group of operator staff. It is, in any case, doubtful that they could have been helpful had they stayed, since most of them were not familiar with the methods and practices of decontamination.

The divisions of nuclear power station staff that are normally assigned decontamination tasks usually only deal with personnel airlocks, laundry, and cleaning of staff rooms with relatively low radiation levels (2000 events/min·cm^2), which does not require special training. The repair team only needs to perform local decontamination in cleaning up after minor jobs, which do not involve high radiation levels. All in all, the number of departments responsible for decontamination within the production union "SoyuzAtomEnergo" is quite small. The best equipped of these is at the Kolsk (WWR) and SRI nuclear power stations.

The tasks assigned to the operational group of the Special Zone were to determine the radioactive levels in the territory and to decontaminate it, all structures, interiors of buildings, and equipment.

It should be emphasized that noone in the operational group of the Special Zone had any previous experience with organizing a decontamination project, certainly not on such a massive scale. No program of recommended action for decontamination was available.

The drawing up of such a program was entrusted to the scientific branches of the Ministries of Energy and Engineering (other academic groups from local research institutes were asked for advice on several occasions). These scientific bodies had available to them a large variety of decontamination technologies and methods which had never before been combined into a single action. The conditions under which they would work most effectively were never established.

4.2.1 The Four Phases of the Rectification

The activities connected with the Rectification of the Consequences of the Accident can be separated into four stages, each with a different primary objective. All efforts were concentrated on containing the radiation released by the exploded Block 4. All production processes were under the direction of the Government Commission.

April 26 – June 1, 1986: The north side of Block 4 and the east side of Block 3 were decontaminated to prepare the area for subsequent work by the Ministry of Coal Industries. Decontamination of the access roads was necessary to allow later work of the Ministry of Defense and the construction division of the Ministry of Energy to proceed according to the time schedule set by the Government Commission. The schedule also defined the necessary contingent of personnel and equipment, but work proceeded without a detailed program or guidelines. The personnel of the power station did not participate at this stage of the project.

At this time, many scientific groups of various organizations with experience in radioactive decontamination visited the area.

The effectiveness of the decontamination was meager: the radiation levels in the decontaminated region did not change. 5000 m^3 of soil were transferred to a hurriedly built "burial vault" within the Zone, however, this vault did not meet any standard requirements for atomic waste storage facilities, even to a first approximation.

The absence of nuclear power station personnel meant that the work was essentially unsupervized. In particular, the access rail serving Blocks 3 and 4 was completely destroyed and other water/power/communications lines to Block 3 were buried under a layer of dry concrete which, along with radioactive soil was wind-blown over the area. The later rectification of these mistakes required enormous effort.

One of the most severe complicating factors in this first phase was the lack of a preliminary plan of organization for rectification of the hypothetical maximum design basis accident (MDBA) of such a power station.

June 1 – July 15, 1986: This phase of the project included the beginning and completion of the preliminary decontamination of Blocks 1 and 2. The basis of the organization of the decontamination was provided by the following documents:

- Program (schedule) of decontamination of the main building and first priority support systems to allow reactivation of Blocks 1 and 2 in October 1986 (approved by the Chair of the Government Commission on June 2, 1986).
- Schedule of priorities and objectives for decontamination of the surrounding territory of the nuclear power station (approved by the Chair of the Government Commission on June 3, 1986).
- Government Commission documents on decontamination.

In accordance with the stipulations of the above, the primary responsibility for the direction of the project was given to a staff of representatives from the Chernobyl nuclear power station, the Building Department of the Chernobyl nuclear power station, the USSR Ministry of Defense, the Ministry of Mechanical Engineering (MinSredMash) and the Ministry of Health, under the direction of the Chernobyl nuclear power station chief designer. However, the appointed members of the staff (except for the Chernobyl nuclear power station and Ministry of Defense personnel) participated in the project only sporadically. In general, all administrative decisions regarding the Special Zone were made in dialogues between the nuclear power station administration and the commander of the operational group of the Special Zone, with daily reassessments at operational meetings.

It was during this second phase that many scientific groups began to make significant contributions; there were plenty of opportunities to make recommendations and proposals. However, these groups did not have a coordinating institution and as a result of the lack of coordination between them, it was very difficult to decide which of the proposed projects was the best. This was the main problem at the second stage.

In addition, there was no supervision by the scientific advisors to monitor the execution of their proposals, and there was no operational analysis. The enthusiasm of these groups faded when the results of their projects were not evaluated and they gradually slipped out of the organization.

Because these scientific groups were neither under the jurisdiction of the Chernobyl nuclear power station administration, nor under the control of the operational group of the Special Zone, these two bodies were also not in an authoritative position to oversee the contributions of these institutions. The primary problem concerning the contribution of the scientific groups was that they did not provide supporting information on the methods that they proposed, such as on the required materials and methods for producing them (for example, regarding chemical reagents).

During this stage, the number of Chernobyl nuclear power station staff members was greatly increased. The staff tried to participate in the decontamination projects; however, because of the need for such well-trained staff, it was necessary to keep the individuals' radiation exposure low so that they would be available to do repair and installation work on the other blocks. For this reason, specialists in decontamination from other power stations, such as the Kolsk Nuclear Power Station were brought in.

At this stage, an administrative body for the Special Zone decontamination project was formed and began active work with the staff of the rectification program. This body drafted several plans to bring in civil and military specialists to organize the operational administration and technical management of the project.

July 15 – November 1986: By this time, the entire administrative structure of the Chernobyl nuclear power station rectification and decontamination project was in operation. The rectification staff decided the following issues and methods of fulfilment:
- decontamination of the Chernobyl nuclear power station interiors;
- decontamination of buildings and other structures in the Special Zone;
- decontamination of the soil of the Special Zone
- dust minimization in the Special Zone;
- minimization of dust coming from Block 4;
- decontamination of the roof of Block 3. The hardest and most dangerous of these tasks was the elimination of the high radiation sources of the roof of Block 3, i.e., in deadly radiation fields of up to 4000 Roentgen/hour.

The specific main tasks in this phase in and outside of the Zone were:
- construction of the containment roof (the Sarcophagus) over Block 4;
- repair and reinitialization of Blocks 1 and 2;
- decontamination of Block 3;
- decontamination of the visitor areas of the Chernobyl nuclear power station, the town of Pripyat, and structures within the 30-km Zone;

- construction of the nearby guard post village "Zelyonyi Mys" for the Chernobyl nuclear power station personnel.

The Government Commission delegated the primary responsibilities for these tasks among the following groups:
- the USSR Ministry of Energy, the Chernobyl nuclear power station and associated agencies were to take care of Blocks 1, 2, and 3, the guard post village, personnel airlocks, and support structures;
- the USSR Ministry of Defense was charged with decontamination.

From December 1986: Phase 4 of the project, which continues to this day, called for the total entombment of Block 4, the repair and powerup of Block 3, the full operation of Blocks 1, 2, and 3, the entombment of Blocks 5 and 6, further decontamination of affected zones, and construction of the town of Slavutich for the Chernobyl nuclear power station personnel.

In November 1986, the production union "Kombinat" was reorganized into one of the major entities running Zone operations. The creation of Kombinat in the form in which it exists today is yet another strategic miscalculation of the rectification program.

Fig. 4.2. The sign "PRIPYAT" used to welcome visitors to that city. Now it is behind the barbed wire that surrounds the evacuated city

4.3 Firsthand Accounts

4.3.1 Radiation Safety in Practice

I spoke with Yevgenii Akimov about the conditions in the Zone which prevailed in the summer of 1986.[2] He came to Chernobyl about a month before I did. From the initiation of the project of constructing the "Sarcophagus" up to its completion, he was the Deputy Chief Designer at Chernobyl nuclear power station. He has been to the most dangerous areas in the Zone and has justifiably earned one of our nation's highest medals of honor.

C: Four years have passed since the eventful summer of 1986, and enough time has elapsed to have reflected on the events of that year and prior. You must have had some new thoughts. What are they, and how and why did you end up in Chernobyl?

A: I've been working in nuclear energy since 1961. If you count my 6 years of education, then I have already been connected to this business for 35 years. I began work at the Siberian Atomic Station as an engineer for reactor operations, went up through the ranks to Shift Supervisor of the reactor block. Afterwards, I transferred to the larger Kursk atomic energy station, where I went from Shift Supervisor to Deputy Chief Engineer of Operations.

C: Were you at the Kursk station from the beginning?

A: Yes. I was there from the beginning. I participated in bringing all four blocks of the Kursk station online.

C: RBMK blocks?

A: Yes. That's why I can say that I really know all about atomic energy production from the inside out. I know its faults, its problems, its victories, and its tragedies.

When the accident at Chernobyl happened on April 26, I didn't know anything about it. Even we, the workers in the atomic energy industry were not informed. The first pieces of information that we got came on April 28, and they were contradictory and confusing. We thought that some kind of hydrogen explosion had occurred, a hydrogen burst. No further details were known and we started to look for possible causes of hydrogen explosions at nuclear power stations. We analyzed all hydrogen sources that we were aware of, but since we didn't have any conclusive information about what happened we could not at all imagine the scale of the accident.

Finally, the contradictory pieces of information started to fall into place. And this despite the strictest secrecy directed even against us specialists.

---

[2] This conversation with Y. Akimov took place in Moscow in April, 1990.

C: Do you mean secrecy even within one and the same department?

A: Even in the same department there was an immense secrecy effort. But even the unclassified data gave me the impression that a large radiation accident had occurred. I know what a radiation accident is from my experience with the last one. I know that in an accident the crew has to be periodically replaced. Going into high radiation fields requires working in shifts. But these crews require competent people who are familiar with the equipment, who know what has to be done.

I considered myself to be a knowledgeable person and so, already on May 2, I realized that they would probably be needing my services.

I called the former head of GlavAtomEnergo, Gennadii Anatolyevich Veretennikov, and told him that I had an idea about the level of danger and what happened, although I didn't know the details. I was ready to participate in the rectification of the consequences of the accident, especially because I was an expert in this area and, moreover, because their station is identical to our station. That is, the complex, the buildings, the layout of the equipment is all familiar to me.

C: Do you mean that there's an exact correspondence between the Kursk plant and the Chernobyl plant?

A: Yes. There's practically a one-to-one correspondence between all four blocks. Of course, every block was an improvement on its predecessor, but these improvements were of a superficial nature.

On May 7, I received orders from Moscow to go. From 6:00 a.m. of May 8 until November 30, I was in Chernobyl, practically never leaving the grounds.

During this time my assigned tasks were rather varied. At first, in May, I was in charge of coordinating the activities of the various work groups that had come to Chernobyl nuclear power station.

C: What was your strongest impression when you arrived on the morning of May 8?

A: The emptiness. It increased as we got closer; I came by car from Kurchatov. After we went through Kiev, and as we got closer, even though we were traveling at night. But I remember the increasing emptiness, the absence of people. I began to feel this with all my senses.

When I entered Chernobyl that morning, I didn't see a single living soul, just a few dogs slinking around. We took the road to Pripyat and finally found someone to ask where the Ministry of Energy headquarters were. They didn't know either, but they sent me to wherever I might see a lot of soldiers and military trucks.

C: Weren't there any reservists around at that time?

A: No.

I did find the troops. Nobody stopped me as I walked into the military headquarters. I finally got the attention of one of the colonels, who directed me to the Ministry of Energy headquarters across the road. It was housed in the same building as the Government Commission.

After we settled in, I went back to the Ministry of Energy offices to report to the Deputy Minister, Shasharin. He told me to have a seat while he finished some business.

There were some quite stormy debates going on in there, but since I was a new arrival, I couldn't really follow what was going on. I did understand, after sitting until noon, that there was quite a bit of discord, that there was no systematic approach being taken to arriving at decisions on fundamental issues.

More and more new problems were coming up with which people were attempting to grapple. A lot of issues popped up suddenly, few had anticipated them. That is, most were completely unprepared for them.

Lunchtime was approaching and Shasharin asked me if I had eaten. When I said, no, he told me to go to eat, I could see for myself what kind of chaos it was. He promised that we'd talk after I got back.

I went to eat. Thank God in those days going to lunch was not difficult. You just walked into any place where you smelled food and noone would ask you who you are and what do you want. They would just feed you and then send you on your way.

C: There weren't any ration cards yet?

A: No, no ration cards yet. The food services were primarily organized by the military. When I returned from lunch, Shasharin spent literally a few minutes with me. He told me, "You see what a muddle we have here. You have an unprejudiced eye, why don't you stay and listen and then write a report. Not really a report, but some kind of little plan of action which could channel some of this energy." I understood.

I started to pay closer attention and to analyze what was happening. That evening I wrote what I thought should be the correct plan of action. I showed it to a number of friends and colleagues and asked them what they thought was right and what was wrong, and if it was wrong, then how to make it right. But everything that I had written they thought was right, too. They only added a few comments. In particular, my former director at the Kursk station, Valerii Kuzmich Galerikhin, was the most helpful – he gave me a few good tips.

I presented the paper to Shasharin the next day. He told me to show it to Il'yin, to have him approve it, because I had so many items to do with medical care. I had no idea who this Il'yin was. He was in a meeting when I went down to see him. Afterwards he received me brusquely, and immediately began to cross out various points on the paper. He was crossing out all the items that had nothing to do with medical matters, like those to do with the Soviet government. He kept saying, "This isn't mine, this isn't ours." Then I said that he didn't have to cross these things out since it was already obvious that they weren't "theirs". There were just a few points having to do with health care and they should be looked at carefully, as to whether they were formulated correctly and ranked appropriately according to their importance.

Later on, it turned out that the proposals that I wrote on health-related issues aroused particular interest. Although frankly, they necessitated a number of complex undertakings.

Some of these proposals he approved for further development, but in the end he just returned my report to me without any constructive changes.

C: Were the medical issues to do with radiation safety?

A: They had to do with radiation safety and also with other health-related measures. In particular, they spelled out the absolute necessity of a health monitoring program, of giving thorough medical examinations first to those that were at the station at that time, second, to those that were in the workers' settlement, third, to those who were evacuated to other regions. In the last group it was necessary to find them first.

To put it briefly, I can tell you that this paper, unfortunately, had no influence whatsoever on subsequent decisions of the Government Commission or on the Ministry of Health Maintenance which was being run at the time by Il'yin.

C: Was he there a long time?

A: I can't tell you how long he was there. I know that he would appear periodically. I think that he was probably in Chernobyl until May 20 or so.

When I reported back to Shasharin, in the confusion he apparently also did not realize the essence of my recommendations. He told me to go to the Civil Defense Offices (we later called it the "bunker") to help out. I asked him what my assignment should be. He said that I should maximize the possibilities of coordinating the efforts of the ministries and departments and other organizations which would later be responsible for the rectification and decontamination of the visitor's area.

So I went there. I worked on these tasks basically all of May. We worked around the clock. There was a constant flow of people. Civilian and military groups which were working on this or that task assigned them by the Government Commission. We had to decide whether it was safe to allow them to work in their assigned area, to ensure safe working conditions, and to arrange electrical supply if necessary. We had to organize their laundry and changing rooms, set up food services, and so forth.

C: How were questions of radiation safety decided? You were in the bunker from May 9. How was it decided how long you could stay and was everyone monitored?

A: It's difficult for me to say whether everyone was monitored, I know that I was not.

When I went to Chernobyl nuclear power station, I had an idea of where I was going and so I had all the means of personal safety with me, among which were monitoring devices. I outfitted myself as well as I could. So, at first I wasn't worried and I didn't pay attention to how other people were being taken care of. But I do know that during this time, not one person came to me to ask whether I had a dosimeter and whether all my readings were alright.

C: Did the medics examine you? When did you have your first blood test?

A: The first time my blood was taken was in August. The medics did not take care of me and there was no analysis of the blood test. That's why Il'yin didn't like my proposals, where I noted exactly these problems which should have been resolved for every single person. Every person should have had a medical examination before and after their work. And each person should have had personal radiation protection gear as well as dosimetric equipment. Without them, it is not only impossible to work under such conditions, it is downright criminal. But people worked.

I can give you an example of the level of dosimetric monitoring that we had; an example I would never have imagined.

One night, I don't remember the exact date, but it was around 3 a.m., while I was at work, a somewhat nervous young man came in. Since everyone wore radiation suits, I couldn't tell who he was. When I was finally done with my phone calls, I asked him whether he wanted to talk to me about something. "Yes," he said. "Go ahead, ask your questions, before the next round of phone calls starts.", I responded.

He said he needed a verification, a certificate confirming where he and his mates had been working.

I said that I was not the one who gives out such certificates, and how can I verify where he'd worked if I hadn't been there and hadn't seen it. But I was interested in how he had phrased the question, so I asked him what he needed the certificate for and to show me on a map where he had worked.

He explained that he needs to give the certificate to the military commander, because without it noone would believe that they had been in areas of high radiation and it wouldn't be put on their records.

I was, of course, completely stunned. I started to listen to him very carefully. We took out a map of the visitor areas and I asked him, "Do you know where you were?" and he says, "Yes." He showed me a point near the Liquid and Solid Waste Store. I saw that we had measured 60 Roentgen/hour there. This was recorded by the chemical troops. I asked him, "How long did you work near this point?" He said, "About 30 minutes."

Suddenly I didn't feel too well.

C: Did they have any dosimetry equipment?

A: They didn't have any equipment.

I asked, "Is that all?" He said, "No there's more." He shows me the area near the pipe bridge. I looked and saw that that point was marked as 35 Roentgen/hour. I asked, "Is that all?", and he answered, "No, there's more." And he pointed to the grounds of the "Evrika" cafeteria. It was later demolished. There there were about 50 Roentgen/hour. I asked, "Is that all?" He said, "That's all. Then we came here to the administration complex."

Thus there were 60, 35, and 50 Roentgen/hour at these places, altogether that's 145, and on average, he spent 30 minutes at each spot. That adds up to 70–75 Roentgen.

These are boys. This was still the regular army, they were draftees doing their normal tour of duty. I asked this boy his name. He was as old as my son. A nice, friendly boy. I don't remember his last name anymore, I only remember that he was called Seryozha.

C: When was this?

A: I think that it was sometime between May 15–20, 1986.

I asked, "But why do you want this certificate from me? Is today the first day that you worked?" "No, the third day," he replied. "Who's been giving you certificates until now?" "Our senior master-sergeant." So I asked, "And where is he now?" The answer was that he's been gone for 2 days. And I said, "So because you haven't been getting your certificates, you decided to come to me?" He answered, "Yes." I asked, "How many of you were there?" He said six. I told him, "Here's some paper, write their complete names exactly, their ranks, the name of your military commander, the number of your division, and then I will write you all certificates."

I knew that if I wrote those certificates, at best these fellows would get a leave, and at worst they might be disciplined. That was the situation in the Army units; in only some, I hope. I can't say anything about all the units, but this occurrence is absolutely shocking. I was literally shaken by it.

I wrote the certificate and he left. A half-hour later he returned with a list of names of the soldiers. I signed the certificate with all my credentials and degrees to make it look more impressive. I never saw this boy again.

The level of monitoring was so low that it astounded me. After all, I was educated in the atomic industry community practically from my childhood, and I never came across anything like this before.

When I was working we had all kinds of lapses in standard procedures. We had accidents with radioactive discharge, but they were all local, within a room, within a building, no bigger, and there the dosimeter reader was the most important person. Without them the foot of the most highly placed official was not allowed to step anywhere. We were thoroughly programed in this.

But here there were no dosimeter readers. At least not enough for the scale of the project. Not only weren't there enough of them, that's only a part of the problem, the real issue wasn't even being raised then. The issue of introducing strict radiation monitoring.

Cleanup work cannot and should not have been conducted until the radiation conditions were fully and accurately known. But they were. Unfortunately, they were.

Our attempts to bring in dosimetrists led nowhere. We tried to get them through the chemical troops. I should give credit to this highly motivated unit of the Soviet Army. Very responsible, and even heroic. But there were not enough of them.

C: Did they have the proper equipment?

A: They had equipment, but its quality left a lot to be desired. And I think that not much has changed since those days in the quality of dosimetric equipment. There just weren't enough people, especially trained workers.

This were the situation and the nature of my work until the end of May 1986."

4.3.2 The Ukrainian Academy of Sciences

> *So far the role of an important learned society, the Ukrainian Academy of Sciences, has not yet been mentioned. Academician Viktor Baryakhtar, Vice-president of the Ukrainian Academy of Sciences, in 1986 the Deputy Chairman of the Operational Commission of the Ukrainian Academy of Sciences on the Chernobyl nuclear power station accident and since 1989 the Chairman of the Ukrainian Academy of Sciences Commission on the rectification program, and Valentin Novikov, doctorate in physical and mathematical sciences, secretary of the Ukrainian Academy of Sciences Section on Physical and Mathematical Sciences and in 1986 the Secretary of the Operational Commission, wrote about the contribution of Ukrainian scientists to the Chernobyl rectification work, the "liquidation of the consequences of the accident (LPA)":*

Today, many documents hidden away behind "seven locks" are becoming the property of glasnost. The curtain of secrecy has been raised on the silencing of the events of Chernobyl. Thus, we can tell about the work of the Academy of Sciences of the Ukrainian SSR in what was then known as the "rectification of the consequences of the Chernobyl nuclear power station accident". Today we have enough information, collected bit by bit, and tested by time.

Of course, it would be good to be able to say that the main job is done and that only a few small details remain. But, alas, it seems as if every year the work begins anew. Bitter experience, gathered a long time ago, hasn't taught anyone anything and is not being used. The authorities of the republic, after prompt evacuation of the inhabitants of Pripyat, Chernobyl, and a number of larger settlements within the 30-km Zone, are following the subsequent events with excessive passivity, and are not actively participating in influencing their course.

Taking all of this into account, it seems necessary to, at least schematically, briefly describe what was done by the Ukrainian Academy of Sciences. And, it would be unforgivable if we forgot the selfless labor of many persons, which has been forgotten and was not built upon.

The Presidium of the Ukrainian Academy of Sciences learned on April 28 that something happened at Chernobyl. The information was very scanty. On April 27, workers of the Institute of Nuclear Studies of the Ukrainian Academy of Sciences during routine government-sponsored tests of the radiation levels in the Kiev area discovered elevated levels of gamma radiation. They followed the track, as it were, and found behind some bushes near the road, a parked automobile and a group of picnickers. All of them and the car were "glowing". The people were immediately taken to the institute and given first aid. These were residents of Pripyat. We would like to point out the interesting closing phrases of the memo concerning them: "After treatment, the victims were directed to the City Epidemiology Clinic." But at the time nobody knew of or was trained for a radiation emergency – and this was already April 28.

Thanks to the existence of the Institute of Nuclear Studies (INS) in Kiev and of other groups of specialists, the period of "radiation illiteracy" was cut short. Yet, we can remember very well how exotic devices like dosimeters seemed to nonspecialists.

The radiation was measured right from open windows. It was rising as we watched: in the course of a few hours it became clear that the situation demanded extreme measures. The same day, in an emergency meeting of specialists at the Institute of Nuclear Studies, it was decided that the first thing that had to be done was to organize the monitoring of radioactivity levels in milk, directly at Kiev dairies. The necessary booths were constructed at the Institute and measurements began already on May 4. Two of the dairies were manned by associates of the Institute of Nuclear Studies, a third by staff from the Institute of Physics, and a fourth by the Institute of Metal Physics.

During the May holidays the organizational work was done, the government of the republic was informed, and on May 3 the Operational Commission of the Ukrainian Academy of Sciences was established, headed by the vice-president of the Ukrainian Academy of Sciences, Viktor Trefimov.

The undisputed leader of the Ukrainian Academy of Sciences Operational Commission was its president Boris Yevgenyevich Paton. His unique skill in making managerial decisions, his knowledge of people and ability to recognize their talents, his authority and impressive industriousness influenced many areas of the work of the Commission and the Academy, and allowed to overcome barriers that would stop a lesser man.

Many people later accused the Academy, especially its president, of causing panic. But we felt that it was best to be prepared for the worst than to close our eyes to that which had a low probability of occurring. After all, the probability of the accident occurring was also very low, but it happened.

The work of the Commission began in an unusual way. We already understood that there was no time to lose and, therefore, we set up the first-order priorities and immediately took the necessary steps to realize them. The standard methods of operation were not applicable, and the possible consequences in terms of some kind of disciplinary action simply didn't cross our minds.

In addition to the organizational duties, the members of the Commission reviewed the subtleties of radiation physics, and reread the now bestselling book *Standards of Radiation Safety* (1976). "Roentgen", "bar", "rad", "Gray", "Sievert", "Bequerel", "Curie" became standard terms in our daily vocabulary.

In order to organize as quickly as possible and increase the rate of collection, analysis, and reporting of data, which could prove pivotal in future decisions, the first step was to move the INS to the Kiev Monitoring and Measurement Station. This allowed it to easily inspect and repair all the measuring instruments. Of course, a large number of instruments was required for a city with a population of several million, primarily, instruments to check radiation levels in milk, bread, and water.

Specialists from the institutes of physics, metal physics, semiconductors, physical chemistry, geochemistry, and mineralogy/geology were involved in data collection. The geologists were the best prepared because in the course of their searches for raw materials they had been confronted often enough with low levels of natural radiation. Thus, they already had the necessary measuring devices.

Only today can we appreciate how incredible the speed was with which this work was done, around the clock, without breaks for sleeping or eating. The first scientists at the dairies were members of the Ukrainian Academy, who not only conducted the analyses, but also trained dairy personnel to use the measuring instruments. In a period of 2 months about 400 persons from various ministries and departments were trained in courses at the Institute of Nuclear Studies. Similar courses were established in mid-May at the Kiev State University.

The Institute of Nuclear Studies and the Ukrainian Academy of Sciences became the primary sources of radiometric information gathering. Thousands of measurements were conducted daily. Unfortunately, the methodology of sample selection, worked out together with the Institute of Nuclear Energy of the Byelorussian Academy of Sciences, was not always conducted correctly, and sometimes it was not even known where the samples came from. For this reason, so much information was contradictory and incomplete that it was impossible to have a realistic picture on any given day of any given territory.

We fully realized the need to systematize the data and, thus, contracted the V. M. Glushkov Ukrainian Academy of Sciences Institute of Cybernetics to do this. Out of this came the now well-known radiation monitoring system. Its first priority was the monitoring of the water supply in the Kiev reservoir and along the entire Dniepr aquifer.

Simultaneously, work on a more general plan was being conducted. In addition to data collection, evaluation, and analysis, we prepared recommendations for the authorities. The primary task at the time was to provide the government with objective information and to stress the consequences of all possible results of the accident, and what we felt were appropriate governmental responses.

The rate at which projects were initiated is exemplified by the following case. The Commission, formed on the evening of May 3, already on the following morning adopted the following agenda:

– to immediately create a monitoring and measuring station at the Ukrainian SSR Academy of Sciences Institute of Nuclear Studies;
– as soon as possible, to prepare for presentation to the governing bodies of the Ukraine information on the evaluation of the worst-case and the best-case scenarios, and to develop proposals for the consequences of the Chernobyl nuclear power station accident rectification;
– by the end of May 1986, to begin preparing a memorandum on the predicted consequences of the accident at Chernobyl nuclear power station, providing for continuous revisions as new data became available;
– before May 5, 1986, to prepare for presentation to the government of the republic, proposals for initial immediate measures to reduce the radiation exposure of the population (standards of behavior, eating, use of medical preparations, etc.)
– to prepare, by May 5, 1986, together with appropriate ministries and departments, recommendations for protection of the food supply;
– to prepare, by May 5, 1986, together with appropriate ministries and departments, recommendations for establishing maximum allowed exposure levels for humans, livestock, and produce.

All of this was subsequently carried out. For example, already on the morning of May 5, the Central Committee of the Ukrainian Communist Party received proposals for formulating instructions for the public about the appropriate behavior under high radiation levels.

It is interesting to note that in the early days of the Commission, when nobody could yet answer the questions "how?" and "what?", an underlying motivation was the further development (which to many, even to leading specialists, appeared unexplainable) of a position of the Ukrainian Academy of Sciences on the undesirability of the construction of the Chernobyl nuclear power station and its expansion, as explained in many letters on many occasions, beginning in the late 1970s.

The next problem that we tackled was that of the water supply in the contaminated territory. Information that the USSR Ministry of Health was preparing new standards of temporary limits of exposure to radioactive pollution, including of water supplies, gave us neither hope nor peace. (These temporary limits were adopted as late as May 30, 1986.) It was clear that this was a political decision. We required a clear picture of what would happen with the waters of the river Dniepr if rains washed radioactive soil into the river. (This work was carried out by the Academy of Sciences, the Ministry of Geology, the Ministry of Water Resources, and the Ministry of the Ukrainian University.)

With Chernobyl, we had the possibility to use data collected over many years of systematic investigations and observations. Before, they were considered by many to be not that necessary for further scientific and technical progress. Already by May 15 the basic parameters for assessing the situation were worked out in detail. After their evaluation at a meeting of the Politburo of the Ukrainian Communist Party on May 20, they were sent to the corresponding executive bodies. Of course, everyone remembers – not only in Kiev, but along the entire Dniepr river – what kind of efforts were made by the government to drill wells and install new water lines. Simultaneously, work was being done to prevent rain clouds from approaching the area of the accident. Today, we can say that we were lucky and the worst case scenario did not materialize, although we were ready to face a situation even worse than the one we had to cope with.

Almost the entire Institute of Colloidal Chemistry and Hydrochemistry of the Ukrainian Academy of Sciences began to work on the decontamination of the water supply directly at the reservoirs, the first priority being assigned to the Kiev reservoir, as it was the closest to the accident site.

Nearly all proposals and recommendations regarding the water supply were put into effect. However, under these conditions, more plans could be made than could be fulfilled. For example, the construction of the "burial wall" around the contaminated zone was not completely finished; fortunately, this wall was not necessary.

A number of other proposals now appear to have been unnecessary, for example, the construction of the Pripyat–Dniepr canal to the north of the Zone, or the absorbing screen on the river Pripyat, since the situation was really not as severe as we had feared; but these projects were carried out in earnest.

The first accomplishment was the measurement of the radiation level in the direct vicinity of Block 4, where the radiation was so high that the instruments available were not able to measure it. New instruments, able to measure 10 000 Roentgen per hour were developed at the Institute of Nuclear Studies. Already on May 10, members of the Institute of Nuclear Studies and the Institute of

Physics made measurements from armored vehicles of the radiation levels near the ruin.

We started to get a clearer picture. The speed with which we could yield visible results (especially on dust reduction) proved the capabilities of the specialists of the scientific establishments, and persuaded many of the correctness of the work of the Academy, thus providing us with moral support.

An important factor which allowed rapid functioning of all entities involved was the departure from established methods of operation, most importantly, the lack of "administrators". Of course, directives from the top were not completely lacking, and they were mostly of the cautionary type: "you can't do this and you can't do that." Sometimes they were simply absurd; how could these people not understand that secrecy was a hindrance? We had to have complete access to reliable information, regardless of whether it concurred with the official version. And who knew what would be declared secret tomorrow?

Particular censure was aroused by our description of the worst-case scenario, until the highest government levels pronounced our work as correct. This did not occur until towards the end of 1986, and not without the participation of Academician V. A. Legasov.

Some ministries of the USSR expressed frank displeasure with the initiative of the Ukrainian Academy of Sciences and accused it of operating outside its jurisdiction. We could not understand how it could be that we, who had the resources to help, should be prohibited from doing so. After all, there were no experts in nuclear accident rectification in our country. We all had to learn by doing, and we were convinced that the scientists of the Ukraine were up to the task.

Consider the situation in Byelorussia. It was worse than it was here, in the Ukraine. In the first few days, there were no real data available. Only in mid-May was the first, fragmentary information received from the Institute of Nuclear Energetics of the Byelorussian Academy of Sciences. Attempts to develop broad-based contacts between the two Academies, initiated at our suggestion, did not go beyond the signing of a protocol on May 26, 1986 by the two presidents of the academies on the necessity of information exchange on the Chernobyl nuclear power station accident.

We also had absolutely no data available from the Russian Republic, and even now, we have no information from them. This is critical because much radioactive mud is washed into the Ukraine, both from Russia and from Byelorussia. At that time, however, we were too busy with our own problems.

Today, looking through the protocols of meetings and private notes, one is surprised at how it could have been possible in that time to even formulate the problems, whose priorities are obvious today. The positions of the Ukrainian Academy on atomic energy, the effects of radioactive contamination on the environment and human health, evacuation and the undesirability of resettlement, the future of the Zone, and the elimination of radioactively polluted material were all verified in the first months after the accident. It is unfortunate that not everything was documented; in those days there was not much time for writing.

Critical situations were unavoidable. For example, we were convinced very early that our equipment was completely outdated. In some cases, the Government Commission could help us, in others the Ukrainian government could. But this was not enough. How happy we were to read the reassurances of Yu. A. Izrael: "Boris Yevgenyevich! You just write what you need. I will personally go to Nikolai Ivanovich [Ryzhkov] and get it." Some of our colleagues mentioned that outstanding foreign equipment was being exhibited at the National Committee on Hydrometrology building. But the equipment was not available to us, so the engineering department of the Institute of Nuclear Studies built new devices. Many perhaps remember the dosimeters at the Kiev city limits, they were built by this engineering department. The prototypes of personal dosimeters also designed at that time have become sufficiently inexpensive and will shortly be put into mass production.

Or take the case of radiation measurements: we all remember what the official numbers were as provided by the Committee of Hydrometrology; but they did not agree with our values. There were even accusations (we do not know by whom) about the "incompetence" of our experts. Academician A. P. Aleksandrov even arranged to send a party of about 100 "for practical support". The first arrivals rather soon realized that their help was not really needed.

A group of scientists at the Institute of Nuclear Studies expressed the need to publish reliable information about radiation levels and were supported by the entire institute. Immediately, there were attempts to persecute this group. The steadfastness of the Academy, to some extent, prompted a review of the contamination levels and the later, second stage of evacuation from the affected territories.

After May 20, some more radical thinkers were already advocating returning the inhabitants to their homes. In formerly classified archives there are, expressed in bureaucratese, the "impressions" of the Ukrainian Academy of Sciences that this cannot be done until it is understood what the consequences of such an action might be. Today, this position is seen to be the only correct one, but then, when the accident was viewed as a temporary event, the position of the Academy was seen as a challenge.

More than 5000 people from 30 institutes and 20 organizations participated in all these tasks. More than 600 of us worked directly in the Zone.

5. The Sarcophagus

"After I had been to the Chernobyl station, I came to the conclusion that the accident was the zenith, the climax of the mismanagement that has plagued our country in the past several decades. . . .

"Even in those days, we were in strangely good spirits. This had nothing to do with our helping with the cleanup work of such a tragic event. Despair was the background in whose light all events were viewed. But enthusiasm was generated by the way in which people worked, by the speed with which our requests were fulfilled, by the efficiency with which different engineering problems were evaluated. We started immediately to evaluate the first designs for the container dome around the destroyed block."

Academician Valerii Legasov
in "Moi dolg rasskazat' ob etom" ["My Duty to Tell About This"] Phantom (Molodaya Gvardiya, Moscow 1989) p. 18, 20

Towards the end of May 1986, the level of radioactivity in the Special Zone had been measured and the work of dropping sand, dolomite, and lead into the ruin of Block 4 (which apparently caused lead poisoning in thousands of people) was largely completed, and it became clear that there was no uncontrollable chain reaction taking place in the reactor. At that time, construction of the Sarcophagus was first discussed at meetings of the government commission.

Block 4 was the main source of radiation pollution. In the time between the start of the initial work and its completion on November 30, 1986, 18 different designs for the optimum containment were proposed. They called for phenomenal quantities of concrete and metal.

Preceding page, construction of the "Sarcophagus", meant to contain the ruined Block 4 reactor, it became a symbol for post-accident Chernobyl.

Of course, these projects also required quite a bit of time. In view of the high radiation it was not possible for humans to approach the site, all work had to be done remotely. Shutting down the remaining three blocks for 2 or even 3 years was not even considered. The priorities of the political leaders made it clear that the containment needed to be installed immediately.

Construction began in June. In the second half of July 1986, the first concrete blocks were installed next to the destroyed wall of the reactor. The Sarcophagus resembles a terraced pyramid. Putting the first layer into place was the most crucial step, as it then acted as an, at least partial, radiation shield, thus facilitating the further work. Towards the end of September the walls of the Sarcophagus grew in giant 12-meter steps to a height of 60 meters.

Three hundred thousand cubic meters of concrete and 6000 tons of metal were used to build the Sarcophagus. Work was conducted under extreme radioactive conditions. Whenever possible, remote control equipment or helicopters were used. When necessary, work was also done by hand, by people who worked selflessly, around the clock.

Only with the help of video cameras, binoculars, and observation from helicopters could the state and stability of the reactor's remains be evaluated. In view of the properties of the soil and the difficulty in assessing the stability of the walls of the ruin, it was decided that the Sarcophagus should be as light as possible. Nevertheless, in July/August it was felt that still more concrete was needed; within 20 days three new cement factories were built within the contaminated territory.

Perhaps one of the most unique parts of the project was the mounting of the scaffolding of the roof: a metal frame 72 meters long, 7 meters wide, and weighing 165 tons was placed on top of the walls; then large diameter pipes were rolled on top of this frame and eventually thin iron plates were put on top of them. It is hard to imagine the difficulty of the task facing the construction workers who needed to get the pieces of the scaffolding into place, to within 15 to 20 cm, without really being able to see what they were doing. Simultaneously, a tunnel was dug under the reactor to make sure that there would be no leakage of core material into the earth.

I discussed various stages and aspects of the construction work with Yevgenii Akimov, the Chernobyl nuclear power station Deputy Chief Engineer.[1] He did a great deal of work on the Sarcophagus and made sure that it was completed in the minimum amount of time and with minimum spread of radiation.

C: When did the actual construction begin?

A: In June 1986. The first thing that we did was to build the retaining wall to hold the poured concrete. It was built on trailers and then moved as closely to the ruin as possible. This was the first step in the construction of the Sarcophagus. It allowed us to come a little closer to the reactor itself, to make the second step. That is, we put up the vertical wall which then acted as a partial radiation shield, and the space between this wall and the reactor was filled with concrete. The next "step" was placed on top of the concrete, a little closer to the reactor, and so forth. That's how we slowly approached the reactor from the side.

The conditions were quite severe – we measured 150, 170, even 180 Roentgen per hour in various places. Using human labor in such conditions is quite difficult.

C: Was the radiation nearer the ruin higher?

A: Of course it was higher. I never went to the ruin myself. I figured that there was no point in it. It was possible to see the reactor from the gallery, and that was close enough to get an idea of the situation. Also, the radiation measured in the gallery was high enough for me.

The construction work proceeded very well indeed. I mean the work on the ground, everything related to the walls, the pouring of the concrete, it was all done with patriotism. The filling of the wall on the side of the generator hall, which was also damaged in the explosion when some scaffolding fell on it, was done in the same spirit. The radiation there was also very high. There were some spots there with 50 to 80 Roentgen per hour. The degree of mechanization used here was the highest in the entire Sarcophagus building project. Platforms were moved into position, they were already fitted with conduits for the concrete, and then the concrete was pumped into them. Between the erected retaining wall and the wall of the generator hall it was poured in with the help of cranes, and then, when it became possible, gravel was poured in with machines. Gravel also buried the transformers. Then it became possible to approach the reactor from the side of the generator hall.

The most spectacular part of the operation was the installing of the metal scaffolding. It was remotely installed using Demag cranes. The stability of

---

[1] This conversation between the author (C) and Y. Akimov (A) took place in Moscow in April, 1990.

this scaffolding determined the stability of everything else. Thank goodness, everything was done as well as it could be done given the possibilities that existed at that time.

C: When was the scaffolding installed?

A: It was installed in late August/early September. I can't remember exactly. Afterwards the work went somewhat faster, when the upper layer of pipes was rolled into place. The surface over the top of the reactor was significantly decreased. Then we could start closing up the side surfaces, too. That work was simpler now, since the radiation was buried under the concrete.

Afterwards, safety barriers were put up - secondary concrete block walls reinforced with lead plates. It was an immense job, an incredible effort. In most cases, especially in interiors, work was done by hand. Building materials were carried in arms. It was too dangerous to put in electrical wiring, because of the possibility of short-circuits. Fires had to be avoided at all costs, because of the completely unpredictable consequences. They could cause another radiation leak which, under the right weather conditions, could spread far beyond the grounds of the plant. That's why so much of the work was done by hand – not because mechanization was impossible, but for safety.

I must single out the people who helped put up the cover on the ruin. All worked with selfless dedication, from those sitting at the drafting tables to those doing the actual building and installing right at the reactor. Most of these people should unquestionably be added to our roster of national heroes.

C: When did the reservists replace the regular army?

A: At the end of May. In mid-May, the directors of the cleanup came to the conclusion that it was a mistake to keep young men, whose whole life was still ahead of them, in these high radiation fields. Immediate replacement of these young men was prevented by some kind of difficulties. But towards the end of May the replacements had begun, and in June there were no more young soldiers in the Zone. They were replaced by reservists between 35 and 45 years of age.

C: During the building of the Sarcophagus there must have been some pretty unusual procedures?

A: One difficulty was erecting the roof onto which the pipes were rolled, and the problem of how it could be made stable enough to bear such a load. It's no exaggeration to say that extremely difficult situations arose daily. It's only because there were so many special situations and because it's been some time now that I can only recall the time when we installed the roof and the last buttresses.

C: What do you mean by buttresses?

A: That's what we called the support between the side wall and the roof. The side shield. We were still working under extreme conditions, in October, when Boris Yevdokimovich Shcherbina, the Chairman of the Government Commission on the Chernobyl Cleanup, arrived and demanded that the shields on the top of the reactor be strengthened. To this day the cover of the Sarcophagus is just a layer of metal plate. The shield itself is just the first layer of pipes under the steel plates which were rolled onto the top of the scaffolding. The concrete that was supposed to go on top was never poured. It wasn't poured for a good reason: because the metal shield was placed directly on top of the old metal scaffolding and it was impossible to determine how strong it still is. So we didn't know how stable the metal cover was either.

So when Shcherbina demanded from the draftspeople that they provide a plan for poured concrete, the head of the design group, Vladimir Aleksandrovich Kurnosov, said, "I will not pour any concrete because the calculations show that that metal scaffolding is unreliable, that it's at its limiting load, and if anyone wants to pour concrete on it they should do it without me." He wouldn't guarantee anything.

Shcherbina was rather displeased, but they stuck with Kurnosov's decision. I think that Kurnosov was absolutely right. In any case, the situation today supports his position, because of the new data that we have. The cover is being continuously monitored and all the investigations show that Kurnosov's decision was right.

C: When the Sarcophagus was being planned, somebody asked what should be done with it after its 30-year lifespan was up. How should it be taken apart? Does the concrete have to be blasted to take it apart?

A: I don't believe that the question was phrased that way, that the reactor cover would last for 30 years.

C: Why?

A: Because the problem was really two-fold. One part was operational – covering the Block, stopping the emission of radioactive particles into the environment. The cover had to be in place in the fastest and simplest way that could be devised in the least amount of time. That's the version of the objective that we kept to.

Getting the cover to last for 30 years or some other length of time was the second problem, which was not really discussed. We thought that once we had solved the first problem, then we could start working on the second.

C: But the Sarcophagus is not, after all, a real tomb? I think that everyone understood that.

A: Of course it's not a tomb. It's not a tomb because the products of the radioactive emissions are distributed beyond the bounds of the first safety barrier. They are not monitored, nor was monitoring included in the project plans. There is also no plan for further cleanup. There is no ventilation system for proper radioactive waste storage. That's why the Sarcophagus can't be called a tomb, although in principle it is a tomb for at least a number of years. It is my deep conviction that there's no point to try to remove the radioactive wastes from the cover. It would cost too much in terms of human lives – the workers who would be exposed to radiation. I believe that under no circumstances should people be exposed to such a danger. Today all our efforts should be focused on preventing all that is inside the Sarcophagus from getting out.

C: Do you think that we are not yet completely ready, technologically speaking, to arrive at a final solution for the Sarcophagus?

A: I have thought a lot about this question. No, we do not yet have the technological means to completely solve this problem. Maybe ideologically we have not yet thought everything through to arrive at a conclusion. We will have to remove large parts (by volume and by weight) of the metal scaffolding. For that we need machinery and methods that do not yet exist, although they may be on the drawing board. We also don't have the equipment with which we could remotely cut the waste pieces to chunks which would be transportable. The solid waste storage facility on site is too small; it couldn't possibly hold all the contaminated pieces. Moving them outside of Chernobyl would not be the right thing to do. New containers have to be built, new means of transferring radioactive wastes from site to site through uncontaminated territory have to be devised.

This is not only technically difficult, but it's also a sociological problem. It would not be justifiable. There's no reason to spread around the pollution that already exists in the Zone. On the contrary, our efforts should be directed towards preventing its spread.

Later, when new buildings have been built, and new techniques developed, then we'll have new ideas, a new theory, and then we can begin to solve these problems with cooler heads.

Today, the more important question is not how to liquidate the pollution in the 30-km Zone, but rather how to deal with the consequences of the accident in the population. The people are very upset, they mistrust not just all things that have to do with atomic energy, but everything to do with nuclear technology. That's why stimulating a further collective psychosis is really uncalled for.

C: Did anybody measure how much radiation was actually released during the Sarcophagus construction project?

A: To take a guess at the actual emission level is simply impossible. I can give you a theoretical guess, but not a practical one. This is simply because, every day, there were new problems and new areas had to be evaluated, spots in which the previous day the radiation levels were still unknown. Therefore, such a calculation was never done.

C: But other calculations were done – so that people wouldn't be exposed to too high radiation doses. The 25 rem exposure level had to be strictly adhered to.

A: That is what we aimed for. The work shifts were planned to make sure that the workers were kept below this level.

But I need to say something quite important: The erection of the Sarcophagus was the responsibility of the Ministry of Engineering and its staff. These are people who in their prior experience had always worked on projects that were more or less dangerous, or involved radiation contamination. Their psychological preparation for this work, even in such an experimental situation, which they could have never imagined, was still much better than that of all other workers, even of the army. They were spiritually and psychologically prepared. Their attitude really helped to move the work along. To put up such a construction, for the first time ever, and in such a short time to solve really complex engineering problems, often without the necessary machinery (with minimum mechanization, but maximum efficiency), is nothing less than an act of heroism.

This kind of spirit was lacking in the other work groups, unfortunately. The speed at which other jobs were completed, even in the high radiation fields, was much slower than we planned. Simply because the workers were psychologically unprepared. It's not the fault of any one person that some jobs didn't get finished, it's simply a question of the workers' previous training. Who would have been prepared for such a tragedy? Nobody. Although we should have been.

C: It was written in the press that the total emissions of Three Mile Island were on the order of 115 rem. Is this per person?

A: Not per person - total. But Three Mile Island and Chernobyl cannot be compared. And secondly, the Americans had a different approach. They simply evacuated the surrounding area and didn't allow anyone in for a sufficiently long time. That is, until the peak activity of the short-lived radionuclides dropped to the maximum allowed levels. Then they went to work.

C: How much did the Sarcophagus cost?

A: I can't tell you anything about the cost, and to be honest, I was never concerned by it. I only know that it was without doubt expensive. Colossally expensive. But I wouldn't put a price on it in roubles. I would value it in terms of human cost, and not just in rem. That's one side of the issue. Then there's the lesson in morality, the psychological lesson which the accident taught not only those who lived near Chernobyl, who worked on the Cleanup, but to the entire world. I believe that these losses are significantly higher than the material losses that were incurred in the construction of the cover for Block 4.

C: Simultaneous with the closing of the Sarcophagus, the roof of the neighboring Block 3 was being cleaned. How were these two projects tied to each other?

A: These projects were indeed coupled. My main job was to coordinate all the work groups. But in addition, I was appointed by the Ministry of Energy to direct the erection of the Sarcophagus; most of my duties involved organizational matters. For example, deciding who could approach the reactor site, which machines should go to Block 3 and which to Block 4 when the two blocks were being separated. It was intended that Block 3 would be brought back on line. Since people were already working under similar conditions in Blocks 1 and 2, maybe it was a reasonable decision to bring also Block 3 back to operation. And since Block 3 is now producing an extra kilowatt, I guess that it was psychologically and economically worthwhile.

But to get back to the question of the connection between the erection of the Sarcophagus and the decontamination of the dome of Block 3 – first we had to clean the roof – all the metal scaffolding, the pipes, even the cooling elements that were thrown onto the roof had to be thrown back off. Most of the debris had collected underneath pipes, which definitely complicated the job. All this garbage was thrown back into the ruin of Block 4, since it was the closest storage container and it also seemed to be the easiest thing to do. But the roof of Block 3 had to be cleaned faster than the time needed for erection of the cover over Block 4.

And it was precisely in this case that our famous bureaucratic chaos in communication stepped in. One department had the task of closing Block 4, and the other had to clean the roof of Block 3. Each of them concerned itself only with its own task without any thought about the other.

I tried to coordinate the efforts of these two departments to the best of my abilities. But for various reasons, and maybe because of the leanings of the departments involved, it was impossible. When the Sarcophagus was already nearing completion, the work on the roof of Block 3 was only in its initial stages. Only the large sized objects had been removed, the small, but radiationally potent debris was still on the roof.

The equipment that they were using was definitely not adequate. The Ministry of Mechanical Engineering was speeding along towards closing Block 4, but the Ministry of Energy, now known as the Ministry of Atomic Energy, was obviously not keeping up.

C: Wasn't it possible to assign the Demag crane temporarily to the cleanup of Block 3? Block 4 wouldn't have exploded, would it?

Fig. 5.1. New equipment is inspected

A: No, Block 4 would not explode, this we already knew. It was completely safe in that respect and there were no additional dangers. But the Demag would have been needed for both the roof cleanup and the construction of the Sarcophagus, and the Sarcophagus had priority.

I think, but this is only my opinion, that this decision was politically motivated. It was necessary to proclaim Block 4 closed and that the danger was past. Even the higher administration levels recognized how tightly coupled these two projects were. But such is the power of our bureaucratic miscoordination, this "powerful force" which allowed one problem to be solved while exacerbating another. This is my personal conviction. I have to blame myself for not insisting, for underestimating my influence, and accepting this decision. In any case, it happened.

C: What can you tell us about the construction of the tunnel under the foundation of Block 4?

A: That was one of the culminating moments in the struggle to neutralize Block 4.

C: This was in May?

A: Digging the tunnel began in earnest about May 21 – 22, 1986. There was some danger and the experts, with their approximate calculations, argued that the rest of the hot core could burn through the foundation and that all the radioactivity would pass into the ground. That's why this danger had to be eliminated.

C: Was there a danger of the reactor, sitting on its cradle, pressing down into earth?

A: Later on we found out that the cradle had indeed not withstood the stress. We know now that the cradle (just two crossed beams) is broken. The lower scaffolding is lying on the floor beneath the reactor. But we didn't know this at the time, since it was impossible to go and look. In any case, we had decided to dig a tunnel under the reactor floor and to put in a safety barrier. This we did in the following way.

In this tunnel we placed heat exchangers. We sunk them in concrete so we could monitor the temperature. If it rose, then we could estimate how close the core was to the ground and then turn on pumps to draw off the heat. Thus, this concrete bed really had two functions – to draw off the excess heat and to prevent radiation leakage into the ground. It was our good luck that this situation never came to pass. That's why it's now safe to debate whether the whole operation was necessary or not. At that time, when we were very uncertain, it definitely seemed necessary. Because, after this job was finished a number of problems were eliminated so we could focus on the remaining ones, of which there were still a great many.

This work was conducted primarily by the Ministry of Coal Industries, and headed by Mikhail Ivanovich Shadov. Shadov complained that the radiation levels were hardly ever checked. Most people in his profession have no acquaintance with such control systems. As much as I could, I helped him. I helped by making sure that before every shift began there was a dosimeter reader on the premises who would check the radiation levels and tell the workers how long they could work. We tried to maintain this practice as much as possible to help them in their difficult job. It wasn't a matter of fear, it was really a matter of conscience. There wasn't any talk about five-fold salaries or such. People sensed that the problem they were faced with could determine the fates of not just the inhabitants of the Special Zone, but also of those beyond it. That's why they worked heroically. Usually by hand, with shovels and wheelbarrows like in olden times. The entrance of the tunnel was where the condensed water was, where the graphite bricks from the reactor were, and the radiation level was up at 200 Roentgen an hour, according to my own data. That's why bringing in too much equipment into the opening of the tunnel didn't seem to be a good idea.

We tried to get people to wear their protective suits. Inside the tunnel itself the conditions were better, although contaminated mud and dust were tracked into it from the outside. Of course, people were exposed to high radiation fields. Especially internal radiation. Breathing and working in respirators under those conditions was so difficult that often it was necessary to really yell at the workers to make them put these things on. Because of these conditions, because people were given a specific job to do which they had to finish within a few minutes, it was hard manual labor. It took a lot of courage, a lot of effort.

C: Do you know what happened to the fellows that dug the tunnel?

A: I don't know what happened to them. But judging from the latest reports in the print media and on television, they must be having a hard time. They're not getting the medical attention they need. In a number of cases, maybe even in the majority of cases, people who took part in the work have all kinds of health problems. Their illnesses are said to be unrelated to their time in Chernobyl and their working in contaminated conditions. This outrages these people. They are very angry, and rightfully so. What outrages them the most is the uncaring attitude of the medical personnel. It's quite surprising. The people that chose the ultimate humanitarian profession, the struggle for the health of mankind, are refusing their responsibility. There is no doubt that they elicit not only displeasure, but frankly, also charges of criminal negligence. These persons who do not take all possible measures to heal these workers are guilty of committing a crime.

C: What lessons for the future has Chernobyl taught us? For example, in the case of another accident, God forbid.

A: One of the worst mistakes that was made was that not enough care was taken with radiation safety and risk minimization. And there is no data bank on all the workers of Chernobyl.

I remember very well a presentation I saw made by a journalist. It was in 1987. The discussion had nothing at all to do with the Chernobyl tragedy. The discussion was about glasnost. One of the journalists made the statement that glasnost is becoming more and more widespread, that one can discuss anything openly. He said that if Chernobyl had occurred 5 years previously then one of two things would have happened: either we would not have known anything about it or, if it had not been possible to cover the accident up, then it would have been presented to us as the best accident in the world. I was completely shaken by this.

I think that underlying all our work projects there was such a chaos, so much was done under the table. The most important question is how safety issues were handled. Thank God, there were no additional accidents, well maybe there were a few exceptions.

I have to say that, in my opinion, the safety monitoring of the workers was very rudimentary. I want to tell you of an example that involves me personally. I was spending 10 to 12 hours right on the grounds of the Chernobyl station in radiation fields that were much higher than at the "Skazochnyi" young pioneer camp where we were staying. But officially, to this day, I am registered as having received only 2 Roentgen. Let me tell you how this came about.

Skazochnyi Camp is a little more than 30 kilometers away from the Chernobyl power station. At this camp the radiation level was measured at between 5 and 12 milli-Roentgen per hour, in different places. So, if I had spent 7 months only on the territory of the camp, never leaving it, then I should already have had over 25 Roentgen. I don't think that there is a more telling example of how the workers' radiation safety was respected.

This was not just my case, of course, but that of many other people as well. I already mentioned how the regular army soldiers were monitored. These are just a few examples of the way in which the responsible institutions worked, in particular, the Ministry of Health. These examples show not only their level of corruption, but also their level of responsibility.

C: And how were the issues of radiation safety resolved in the atomic energy industry? You said that the primary means of control was simply dosimeter measurements.

A: Actually, in the atomic energy industry, as paradoxical as it may seem in light of Chernobyl, issues of radiation safety are decided at the highest administrative levels, as far as I know, just as in so many other areas of our economy. There are a number of barriers. There is an automatic radiation monitoring system, a system of individualized monitoring and a strict security system regulating admittance. A trained dosimeter reader measures the radiation levels and determines the time that a person should be allowed to work. In all my 30 years of working in the (nuclear) industry, I have never heard of anyone being exposed to more than 5 Roentgen per year. If there were such cases, then they were related to accident containment, and then the workers knew that they were going into areas of higher radiation. But even these doses, as far as I can remember, never exceeded more that 10 to 15 Roentgen a year. And the workers also received compensation. So the radiation monitoring was organized at a sufficiently high level with the necessary trained personnel.

But then the accident happened. There weren't enough trained dosimetrists. There weren't even enough dosimeters. They were rushed in, but the models that were supplied were reliable only above 0.5 Roentgen. If the readings were less than 0.5 Roentgen, they'd just write 0. So if a person spent a whole month working in a field of 0.4 Roentgen he'd still be registered as having a zero dose. I guess the same thing happened to me.

I have no idea how this happened. I learned about this only much later, maybe in September. I was never particularly interested – I figured that I was being monitored, but that they were simply not reporting the results. I was used to the idea of secret data and accepted that they would not be made public.

C: Why should the dosimetry data be secret?

A: I don't see the logic in that either. Before, when I was constantly living in conditions where this or that was always being declared classified, I just took it for granted that dose levels would also be secret. I didn't think about these things very much. We knew that somebody was monitoring our exposure, and that if it became necessary, they would warn us.

C: But sometimes a human life can depend on these data.

A: Yes, but in routine work there are radiation monitors and people who will look out for you. That's why I never had the idea that someone would lie to me.

C: Were you perhaps just unlucky that your exposure was not monitored? Were your coworkers monitored?

A: No. As I later found out, the same thing was happening to other people, too. My friends had already persuaded me to at least go have a blood test. This I did for the first time in August. When I tried to find out the results of the test, I was told that they weren't ready yet. Then I got so busy that I forgot to ask again. I still don't know the results of that test. This is the kind of "monitoring" we had.

They weren't supposed to just monitor, either. They were supposed to take some kind of preventative measures as well, like taking people out when they had reached their maximum exposure. The maximum exposure had been set to 25 Roentgen. But I didn't see any organized efforts to protect the safety of the workers of the Chernobyl Cleanup.

6. In the Jaws of Hell

> "What! Of course there was no panic!
> What do you mean? Would I lie?
> If we are called, are we not always glad?
> Ready for anything – even to die?"
> *Vladimir Shovkoshitnyi*

It was September 1986. Public dis-information reached its highest level. The attention of the whole world was focused on the Special (10-km) Zone. Many people had already died and were buried in lead graves in the Mitinsk cemetery near Moscow. The press reported that we had managed to overcome the consequences of the disaster very speedily, yet, it was the time when we were just entering the high radiation fields: 1000, 2000, ... 4000 rem per hour. Such were the radiation levels that we encountered in early September 1986 on the roof of Block 3.

Then it should have been the time to stop, to wait, think, and prepare special routines and procedures to tackle the problems in these deadly conditions. But no! "Press on, bring the remaining three blocks of the damaged power station back to production quickly!" – those were our orders.

But work came to a halt. German, Japanese, Soviet robots – none of them were designed to endure such superhigh radiation fields. The solution was simple: allow a one-time dose of 20 rem per person and send soldiers onto the roof. "Biorobots".

On September 19, 1986, people were sent into the jaws of hell.

Preceding page, the bitterly named "biorobots" at work, dressed in futuristic protective gear, which offered only inadequate protection against the lethal levels of radiation. (Photo taken in September 1986.)

6.1 The Roof of Block 3

In the most dangerous parts of the post-accident Chernobyl site, accident radiation levels exceeded 500 Roentgen/hour. The area around Block 4 was one such zone – the level there was 400 Roentgen/hour; others were zones "N" and "M", the rooftops and the structures of Block 3, where radiation levels reached 800 to 1000 Roentgen/hour or more, and the chimney platforms.

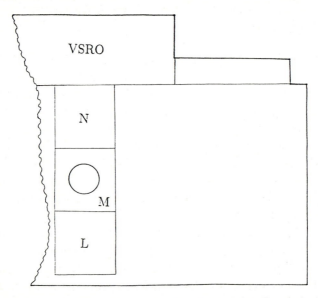

Fig. 6.1. A sketch of the roof of Block 3 with its high-radiation fields. The circle denotes the base of the ventilation chimney (shown in the photograph at the beginning of Chapter 5). L, M, N, refer to the different zones described in the text; VSRO is a Russian abbreviation referring to the installation in which liquid radioactive waste is dealt with.

Zone N is the roof of building 6001 of the ventilator house. The zone measures 24 × 24 meters. The height of the zone is 61 meters. Around the perimeter of the roof is a raised edge 0.4 meters high. The covering of the roof is made of ferroconcrete slabs measuring 585 × 100 cm and weighing 1 – 3 tons. On top of these is a bituminous layer 5- to 10-mm thick. A pipe 150 mm in diameter runs around the perimeter of the zone.

Zone N is right up against the side of the vertical wall of building 7001. Protruding across part of the zone through this wall is an air pipe 1000 mm in diameter. Zone N is the middle step in the roof shared by Blocks 3 and 4.

The explosion in Block 4 broke through the ferroconcrete roofing slabs of the central reactor hall and hurled them in all directions. Some landed on zones N and M. The components of the reactor – partly destroyed fuel rod assemblies, fuel rods, graphite moderator blocks of a total weight of between 140 and 150 tons were also scattered across the roofs of these zones.

Radiation reconnaissance scouts brought back estimates of more than 20 tons of debris in zone N, about 100 tons in zone M, and 15 to 20 tons on the chimney platforms.

The radiation scouts from the "Atomic Energy Repair Union", A. S. Yurchenko, V. M. Starodumov, and G. P. Dmitrov, were the first to go out into the high-radiation zone and monitor the level of ionizing radiation in zone N. The levels ranged from 100 Roentgen/hour at the entrance, to 900 Roentgen/hour around the perimeter.

B. Ye. Shcherbina, Deputy Chairman of the USSR Council of Ministers signed the authorization implementing the decision of the Government Commission on August 19, 1986 to hand over to the USSR Ministry of Defense together with the power station management, the task of clearing the highly radioactive debris from zones N and M and from the chimney platforms, and dumping it into the ruins of Block 4.

Earlier, a plan had been worked out for the removal of the debris by "Demag" cranes. However, this plan was abandoned when all the Demag cranes had to be transferred to speed up construction of the sarcophagus around Block 4. There were attempts to use both Soviet and foreign robots, but these all broke down when put to work in areas where radiation levels were high.

The task of shifting the debris – the broken fuel rod assemblies, the rubble etc. – was assigned to men equipped with scrapers, tongs, shovels, hooks and stretchers. The plan was to collect the debris, shift it to Block 4 and dump it all into the ruins of the shattered reactor. Incoming personnel, organized into units and sub-units, were directed to buildings 5001 and 6001, where each unit was entrusted with a particular task; they were given a general briefing about how the work should be carried out. Briefing of officers put in charge of units was carried out on the evening before each operation. Officers serving as guides and pathfinders were given a special briefing.

The work of throwing all the radioactive debris into the crater of Block 4 was rather problematic, since the routes that had to be taken from the initial points to the reactor were very difficult and dangerous.

The management of the troops engaged in the work was regarded as a matter of the utmost importance. For this reason a special command post was erected near building 6001. Television monitors were installed there. The command post was also equipped with short-wave radio, linking it with the rectification control center and military headquarters. Greatly enlarged photographs of the most dangerous zones (N, M, and the chimney platforms), plans showing routes into the zones, and models of fuel rods, fuel-rod assemblies, graphite moderator blocks and other dangerous objects were provided.

The personnel who arrived to carry out the work (enlisted men, non-commissioned officers) were volunteers. The unit and sub-unit commanders selected the men and gave an introductory briefing. Each man signed a certification that he had read and understood the basic rules – especially important were the rules governing the use of safety equipment.

The scope of the various tasks was established daily by the Ministry of Defense operations group, the scientific center and the Chernobyl rectification control center staff. Orders were then conveyed to unit and sub-unit commanders not later than 12 hours before the work was due to start.

Initial briefing of the commanders was carried out by representatives from the Scientific Center and the Rectification Control Center staff. Personnel were prepared for their tasks with the help of a specially constructed model of the site which reproduced the current situation in zone N and made it possible to acquaint them with the methods being used.

The following protective gear was issued:
– to protect the lungs: respirators of type "Astra-2", "RM-2" or "Astra-1";
– to protect the spinal cord: lead sheets 3-mm thick;
– to protect the testicles: a lead truss 3-mm thick;
– to protect the eyes from dust: goggles with glass no less than 2 mm thick;
– to protect the face and eyes from β-radiation: an acrylic plastic mask 2 – 5 mm thick;
– to protect the feet: lead insoles 1.5-mm thick to be worn inside the boots;
– to protect the body (chest and back): rubber aprons lined with lead;
– to protect the hands: cotton gloves, over which were worn lead-lined mittens;
– to protect the neck and throat: a lead sheet at least 1.5-mm thick.

The protective clothing was fitted and adjusted in a specially equipped area by the commanders of the units and sub-units. As a rule, a battalion commander and his staff were always present.

In order to monitor the dose received by the personnel a special post was set up where radiation intelligence officers and unit commanders checked dosimeters. These dosimeters were issued to all military personnel and worn under the chest protector. Readings were entered in a register.

While one team was at work the next team would be waiting, already fitted out and alert, ready to render assistance if necessary. Throughout the operation constant vigilance was maintained in order to ensure the safety of the men working in the most dangerous zones.

In all the time when men were working in these zones there was only one incident – when a soldier who was throwing a block of graphite into the reactor twisted his ankle and fell. The shift commander and the reconnaissance officer immediately went to his assistance and led him away from the area. One case is also recorded when a soldier at a briefing, watching the work in progress in the zone via a television monitor, fainted from anxiety. He received medical help and was sent back to his unit.

When a team arrived at a work site, 10 – 15 seconds were set aside for a look around the zone and at the disposal area, i.e., the ruins of the reactor. The norm of the amount of debris to be moved by each man was worked out as 50 kg of graphite or 10 to 15 kg of fuel rod fragments.

The average time spent in the zone was from 1.5 to 3.5 min. The average radiation dose received was 12 to 20 rem. The clearing of the high-level radioactive debris from the roofs was accomplished in the period September 20 – 30. The operation carried out by these Soviet troops, who cleaned up areas where radiation levels reached 500 Roentgen/hour and more, has no parallel in world experience with nuclear radiation.

6.2 Clean the Chimney Platforms!

Work began also on the cleaning-up of chimney platforms 2 to 5, the idea being to reduce the amount of radiation affecting zone M. When the work was completed the radiation level on the first chimney platform was no more than 30 Roentgen/hour. Inevitably, the cleaning of the platform involved throwing debris down into zone M, from where it was thrown into the ruins of Block 4. The radiation levels in zone M and on the chimney platform were reduced by a factor of 10, which made the organization of the subsequent operations somewhat simpler. However, work at great heights called for fit and courageous men. Age was a secondary consideration. The soldiers who worked in zones N and M and on the chimney platform were in the 35 to 40 age group or older.

The Government Commission decided that the task of clearing the chimney platforms should be entrusted to trainees from the Ministry of Internal Affairs fire brigade colleges at Kharkov and Lvov. Volunteers came forward from among the trainees and they were selected on the basis of their fitness as, displayed in work-related sports, and on their ability to work at great heights on lightweight ladders.

On September 30 the trainees arrived at Chernobyl to be assigned their tasks. At 17:00 hours on the same day an experiment was carried out to establish the time taken to climb to platforms 2, 3, 4, and 5 and then return to the starting point. At the same time, the volunteers were to estimate the amount of work awaiting them on each platform and also monitor the radioactivity levels.

Trainee Viktor Sorokin managed to carry out this initial task in 15 minutes, receiving a dose of 18 rem in the process. The experiment made it possible to calculate how long the work on the chimney platform would last.

A program was drawn up for the decontamination of the platforms:

- platform No.2: 10 minutes and 10 men;
- platform No.3: 15 minutes and 6 men;
- platform No.4: 20 minutes and 2 men;
- platform No.5: 25 minutes and 2 men.

October 1, 1986: At 09:00 hours the trainees arrived at Building 5001, where they were told the results of the experiment and told what their tasks would be. Kharkov fire brigade trainees V. S. Gorbanko and P. K. Kushakov were the first to set to work. They climbed to platform No.5 and cleared it completely. On the way down they decided to go beyond the call of duty and do some work on platform No.2. However, they were called back to the start line by a siren. Altogether this pair had been working for 22 minutes.

A second pair was ready to tackle platform No.4; they were trainees A. I. Florov and V. M. Zubarev. It took them 20 minutes to clear their platform.

Teams were sent at 15-minute intervals to platform No.3. Trainees V. G. Kosogorov, A. V. Kotsyuba, Yu. Lobov, V. V. Lunonets and – from the Lvov college – I. D. Blashko and A. V. Svetitskii took 1.5 hours to clear platform No.3 completely.

Teams were sent at 10-minute intervals to platform No.2. The team members were Lvov trainees N. S. Pridius, Yu. V. Saulen, A. P. Dremlyuga, Yu. S. Kolachun, S. N. Klimchuk, V. S. Ilyin, and two Kharkov trainees – V. S. Michkinevich and A. V. Goshev. Major M. V. Sudnitsyn and trainee V. N. Avramenko helped in the clearing of this platform. A total of 1.5 tons of radioactive debris was cleared from chimney platforms 2, 3, 4, and 5. The lumps of graphite which had been thrown down from the chimney platforms were cleared from zone M by soldiers of the 258th Mechanized Civil Defense regiment. Only 46 men were involved.

The cleanup operation in the special danger zones and on the roofs of Block 3 painfully showed that neither the Soviet Union, nor any other country had the techniques or the technology needed to deal with such a task. In practice, the emergency work – the clean-up of the graphite, fuel rods, etc. – was done with small, improvised tools and devices. The robots broke down when their electronic equipment was exposed to high-level radiation. No means were available that could provide effective protection for people working in zones where radiation levels reached and exceeded 500 Roentgen/hour.

In the period of preparation before the work started, protective outfits were designed and made which reduced the absorbed dose by a factor of 1.6. This outfit could be the basis for a more sophisticated version which would give better protection and comfort while hampering the wearer less. Because of the effect of ionizing radiation on the whole body, it would be appropriate to develop a one-piece, coverall-type suit.

One of the problems was the difficult access to the work site, for humans and robots. Future development of robots should take the following factors into account:

- the effect of high levels of radiation;
- the inaccessibility of many work locations;
- the complexity of routes taken through the work area, and the maneuvering required, often in high-radiation zones;
- the unevenness of the terrain likely to be encountered (including heaps of graphite, fuel rods, etc.);
- the probable need to move and dispose of lumps of debris weighing up to 3 tons;
- the desirability of remote control from a safe place.

6.3 A General Recalls Details of the Battle

Major-general Nikolai Tarakanov of the USSR Defense Ministry, supervisor of the decontamination operation on the roofs of Block 3, gives in his recollections [6.1] the following account:
The most demanding and dangerous decontamination job was on the roofs of Block 3, where a significant quantity of the highly radioactive debris from Block 4 landed. There were pieces of the reactor's graphite moderator blocks, fuel rod assemblies, zirconium tubes, and so on. The doses given off by the various objects lying on the roofs were extremely high and threatened the life of anyone handling them.

From April 26 to September 17, 1986, this enormous mass of material lay on the roofs of Block 3 and on the platforms of the main ventilation chimney. It was blown about by the wind and washed down by the rain as we prepared to shift it. We had great hopes for the success of the robots that we had been promised. Finally they came. Several robots were lifted into the special danger zones by helicopters – but they would not work. Their batteries were soon dead and their electronics fouled up. Somewhat earlier a plan had been prepared: "The Decontamination of the Roofs of the Main Block and of the Ancillary Buildings". This plan had been drawn up by a certain Moscow institute. It envisaged the use of two Demag cranes (made in West Germany, they cost 4.5 million roubles) to clear the roofs with adhesive grabbers, to use hydraulic giants and pumps working at a pressure of 8 to 10 atmospheres, and the use of "Vorstern-770" hydraulic manipulators (made in Finland). In addition, it would be necessary to build special concrete roads for the Demags to navigate.

However, practically no dosimetric or engineering reconnaissance had been carried out before the plan was drawn up (apart from some photographs taken from helicopters). The supervisor of the radiation reconnaissance staff Aleksandr Yurchenko had made some sorties. The first time he went to zone N the reading on his DP-5 monitor went off scale. This brought the reconnaissance sorties to a halt for a while. The plan was to measure the radiation levels in the accessible places by means of a dosimeter attached to the hook of a crane.

In order to minimize the shifting of debris from place to place and also to avoid the need for special disposal pits, it was decided that the debris should be dumped into the ruins of Block 4. Therefore, the roof of Block 3 had to be cleared before the completion of the sarcophagus, but constructing the latter was the primary task and using the Demags for roof-cleaning work would have delayed its completion so this plan was not carried out.

It was essential to find an alternative which would resolve the clash of priorities. Another, more detailed reconnaissance was carried out by members of comrade Yurchenko's special radiation reconnaissance unit. The roofs were divided into zones which were given the designations N, M, K, etc.. The borders of the zones were the various height levels and the walls of the superstructures. A similar radiation survey was carried out under the roofs of Block 3.

Acting on the new information concerning the roofs of Block 3, the rectification control center proposed another cleanup program. This would involve extensive use of a Demag crane which was to be put in place at the north end of the block. The plan also envisaged the use of hydraulic giants, remote-controlled machines, and semi-remote manual devices. A "Liebherr" crane would also be needed to place the equipment on the roofs. The plan was approved by the government commission, but was also later abandoned because of the unavailability of the Demags.

Throughout the operation that I directed in the special danger zones, I never once saw a robot actually working, although I saw one which had to be dragged out of the graphite. It was burned out by the radiation and had become an obstacle to the work in zone M.

By the end of September, the sarcophagus (this vivid word had become its usual name) was ready to be closed up with large-diameter metal pipes. This job presented enough difficulties of its own, but was further complicated by the need to clear tons of radioactive debris from the roofs and chimney platforms before it could be completed. Whatever the cost, this debris had to be collected and thrown into the gaping hole of the ruined reactor, and then covered with a solid roof. If this was not done the removal of the debris to safe disposal areas would drag on for months.

But how were we to tackle those zones of deadly radiation? Attempts to use hydraulic giants and other mechanical devices proved unsuccessful. The press had made a fuss about the "magic" robots, but they had failed. In addition, the places where the debris had landed, up against the ventilation chimney of the main block and on the chimney platforms, were inaccessible. The height of the structures in question ranged from 71 to 150 meters. It became evident that sending in men to do the job was the only possible solution. That was the conclusion reached by many experts and members of the government commission.

On September 16, 1986, following the instructions I had received in a coded message from Lieutenant-general B. A. Plyshevskii and General Yu. M. Vaulin at Chernobyl, I flew there by helicopter to attend a meeting of the Government Commission. The topic under discussion was the decontamination of the roofs of Block 3 and of the chimney platforms. I arrived at 16:00 hours and reported to Plyshevskii. We went at once to the meeting, which was chaired by Shcherbina in his office. Yurii Nikolayevich Samoilenko delivered a report on the matter at hand. We went over to the relief map on which red flags and other symbols marked the radiation pattern. He gave us a clear explanation of the situation, calling our attention to the special danger zones. Samoilenko reported that all attempts to remove the debris by mechanical means had come to nothing. Only one course of action remained – to call in soldiers to do the job manually, with the simplest mechanical aids.

There was an oppressive silence. Each one of us knew what the danger would be for those who did this job. That was the first thought. Secondly, was this really what a century of technological progress had brought us to? In the hour of direst need we had neither the technology nor the techniques to deal with a job of this kind! Shcherbina went once again through all the other options – not one of them offered any real solution. Then the chairman of the Commission turned to Plyshevskii and said: "I shall sign the government order which will bring in the Army." Plyshevskii replied: "The troops must have orders from the Ministry of Defense." Shcherbina promised to personally contact the Ministry of Defense, and said that we must prepare for the operation.

There was unanimous acceptance of the decision. It really was not taken lightly – but there was no other way. It was also decided that I would be in charge of the operation at the scientific and practical level. At the same meeting there were many suggestions as to preparations for the operation, including a proposal for a preparatory experiment. The proposal was accepted. On September 17, a helicopter took us to the site chosen for the experiment, i.e., zone M. Aleksandr Alekseyevich Saleyev, Candidate of Medical Sciences and Lieutenant-Colonel in the Army Medical Corps had been tasked with testing the possibility of working in the danger zone. All radiation safety precautions were taken. Saleyev was dressed in protective gear made mainly of lead.[1] All this gear managed, as the experiment showed, to reduce the effect of the radiation by a factor of 1.6. On top of all this he had to carry about ten instruments and monitors of various kinds. He followed a carefully worked-out route. His task was to gain access to the roof via a hole in a wall, inspect the rooftop, take a

---

[1] This protective gear has been described earlier.

look at the ruined reactor, throw in five to six shovelfuls of radioactive graphite, and then return when he heard the signal to do so. Lieutenant-Colonel Saleyev managed to carry out this sequence in 1 minute 13 seconds. We watched his every move with bated breath.

We presented a report on the experiment, along with our own conclusions, to the Commission. The members of the Commission considered the report and the documents we had prepared (instruction manuals, memoranda, etc.) for the military workers and gave their approval.

In our recommendations we included a list of qualities required in the volunteers for this job. They had to be psychologically tough, able to adapt quickly. They also had to have the physical strength to carry out their tasks in an extremely short time, before they reached their their permissible exposure dose. The following principles were obeyed in selecting and training the soldiers for these tasks: the first qualification was a willingness to do the job under the ultra-extreme conditions that prevailed; secondly, along with the primary medical screening there was a further selection process which sought out capable, precise, calm, balanced, observant men; thirdly, the volunteer had to be physically strong, fit, and well coordinated. The age limits were 30 to 45 years.

The selection process was carried out by experts who looked for this ideal combination of personal qualities. This process gave us workers who could accomplish the tasks and use up dose limits with the maximum possible efficiency.

However, throughout the whole period from July to November 1986, while the rectification control center was in charge of operations, the USSR Ministry of Health never came forward with any recommendations and did not carry out any examination of the workers regarding their psychological state. Even physiological monitoring was rudimentary.

The members of the special radiation reconnaissance unit had their blood tested only once in 4 months of working in high and super-high radiation areas. The state of affairs in some other departments was no better than that of the medical service.

Great importance was attached to training activities. In September 1986 a training area was set aside on the site where the fifth and sixth reactor blocks were being built. It was provided with dummy graphite blocks, fuel rod assemblies, zirconium tubes and realistic rubble, as well as access holes and pathways like those in the danger zones.

From September 18, a mock-up of zone M was available for training the volunteers. A rough idea of the scale of the task was obtained by aerial reconnaissance. The photographs helped us to consider our strategy and prepare a plan of action.

In order to provide access to zone N a 1-m opening was cut through the side of the ventilation duct. Later it was enlarged to 2 m × 1.5 m.

The preparations continued at full speed. Soldiers made protective outfits by hand. Each outfit weighed 20 – 25 kg.

By September 18, the feverish preparations were completed. Everything had to be done hurriedly because of the time lost while the Commission was still pinning its hopes on robots. The method which we were eventually forced to adopt had not been given sufficient consideration so that, in the end, we had to rush to devise and make simple mechanical aids and protective outfits in the short time available.

The Ministry of Defense appointed a special commission to monitor working conditions and supervise the troops and their support structure. General I. A. Gerasimov was appointed chairman of the commission. He was to lead the troops in the first, most difficult days of the Chernobyl epic. The commission was made up of representatives from all branches of the services. Rear-Admiral V. A. Vladimirov represented the chemical troops. There were also representatives from the General Staff, the Chief Political Bureau, the engineers, logistics people, etc.. After a meeting with the chairman of the Government Commission, Shcherbina, we all went to the airport, boarded a helicopter and took off to survey the scene of operations.

The helicopter first hovered over Block 3, then right next to the main ventilation chimney, then by the ruined reactor itself. All the ejected debris would have to be put back to from where it had come from. It looked as if the total weight of debris that would have to be shifted was more than 100 tons – and it would all have to be collected, carried, and thrown into the ruins by hand.

The Commission returned to Chernobyl to discuss the upcoming operation once again. At this meeting only military men were present. Hardly anyone wanted to speak. Only Rear-Admiral V. A. Vladimirov made a suggestion – that the maximum permissible dose should be increased.

Gerasimov reported on our state of preparedness and on the situation as a whole. He said very firmly that, apart from the Army, no-one would be able to tackle this job. He received the go-ahead.

Within half an hour I was at my command post in Block 3, where the soldiers had been waiting impatiently.

On the afternoon of September 19, the chemical troops, under battalion commander Major V. Biba, dressed in their protective gear were ready and my assistants briefed them. They were informed of the Ministry of Defense's decision that this operation would only be entrusted to the men of the Soviet Army. I said that any man who was ill, or did not feel well, should leave the ranks. The ranks did not stir. That is how it was every day, throughout the operation.

I shall name the first five men who started the difficult operation in zone N. They were battalion commander Major V. Biba, section leader Sergeant V. Kanareikin, Privates N. Dudin, S. Novozhilov, and V. Shanin.

On the command "Go!" the stopwatch was started. The men went out into the zone and set to work with grabbers, scrapers, and shovels. In the course of this operation there were occasions when we came near to abandoning it. The difficulties of the task seemed to be insurmountable and it would have been criminal to continue exposing people to the radiation.

Each soldier went just once into the danger zone and only for a very short time.

I will always remember October 1 – the final day of our operation. There was a particularly large amount of work to do. There were two damaged robots in zone M, they had been mired in the piles of graphite and other debris. We managed to pull them clear with the help of a helicopter, but in order to clean them off, disentangle them, and secure them we had to send in several shifts of soldiers.

Then the hydraulic giants and high-pressure hoses went to work, under the direction of Viktor Golubev. There was no end to the courage and technical resourcefulness of this man.

At 20:30 hours a team of chemical troops made up of Sergeant V. Parfenis, Privates V. Borisovich, S. Mikheyev, and Ya. Tumanis threw the last pieces of graphite and the last fuel rod fragments into the ruins. The siren gave a longer wail than usual. Everyone in the command post shouted "Hurrah!" It was decided that the flag would be raised on the ventilation chimney the next day.

All our command post staff climbed up a fire ladder and through a hole in the roof out into zone M. A short meeting was held at which we shook hands warmly with Yurchenko, Sotnikov, and Starodumov. They were entrusted with the raising of the flag. They climbed up with the flag, hoisted it, and came back down without mishap. We congratulated them warmly, embraced all who had served with us, then took a big panoramic photograph of "our bridgehead".

Thus ended the first phase of the decontamination of the roofs of Block 3.

Fig. 6.2. Ready to hoist the flag

7. The Rectifiers: Then and Now

"Any man's death diminishes me
because I am involved in mankind.
And therefore never send to know
for whom the bell tolls –
it tolls for thee."
John Donne
"Devotions"

During the rest of 1986 and the first 6 months of 1987, work proceeded in the superhigh radiation fields of the Zone. So-called technical decontamination, which has since continued in the Zone for nearly 5 years, did not result in a safer environment.

The emergency teams who worked in the Zone until the end of May 1986 were allowed to return home. The reservists went back to their normal jobs.

During 1986–87, the press called those who took part in the Chernobyl cleanup "heroes". They numbered more than 650 000. Now they are called "liquidators" or "rectifiers" and have been forgotten. Their time spent in the Zone, working in superhigh fields without adequate protective gear and medical attention did not leave them unscathed. They began to fall sick, become invalids, many retired from life.

In a last effort to call attention to the neglect of their social needs, some of these already afflicted people went on hunger strikes and founded the "Chernobyl Union".

Preceding page, construction work on the tunnel dug under Block 4. (Photo taken in September 1986.)

7.1 A Rectifier Speaks Out

I. I. Sgorelo, 51 years old, Candidate of Technical Sciences, now a Group-1 invalid suffering from "ordinary illness" [1]

Words like "compassion", "social justice" are repeated more and more often here in the USSR. For thousands of people, they sound like the crudest mockery of them and their fate. The peaceful atom of Chernobyl went on the rampage and the resulting storm struck these people. They are the ones who answered the call, the ones who, in a cotton suit and a thin gauze mask over nose and mouth, did their best to wipe out the mistakes and miscalculations of others, to lift some of the threat that was hanging over huge areas of the country. The top people in the medical profession and the representatives of the Ministry of Atomic Energy reassured them with soothing stories about the harmlessness of the doses they were receiving.

Time passed and these people are practically forgotten. Even in the Congress of People's Deputies not a word has been said about them. And yet, this was a war against an enemy as cruel and cunning as you can imagine. For many of the veterans of this war the consequences are only now beginning to become clear.

On a television program linking the USSR and the USA the American experts estimated our country's possible losses following the Chernobyl accident as 68 000 people. Yet our press reports the deaths of a dozen people at the power station when the accident happened. Then there was a fleeting reference to the deaths of the firemen who were the first to bear the blow. But what about the "rectifiers" who often worked in the same cruel conditions? Because they are scattered across the whole country and cloaked in the utmost secrecy, all kinds of bright pictures can be painted concerning their health.

Who can, realistically, look after the rectifiers' welfare? That is, who can provide a qualified medical service, pay an increased disability pension of not less than 70% of salary, so as to give them a normal life and the chance to get treatment, in many cases to bring up children? Who can allocate them places in sanatoria, if that is what they need?

Many remember that, for example, in India, after the great chemical plant disaster in Bhopal, the government acted as the defender of the victims. The Soviet authorities, however, the Party and Trade Union organs (at the provincial level) have turned their backs on the rectifiers. The organizations and agencies who sent these people to Chernobyl clearly think that their duty was discharged when the government decided to pay "coffin money", as the rectifiers call it, for the time spent working in the danger zone. But was it these people's fault?

---

[1] From a letter to the Chernobyl Union of December, 1989. To protect the anonymity of the writer, his name has been changed. This letter and the others reproduced later were given to the author by the Chernobyl Union to be used in this book.

Or was it that of the designer's, the ministry's? The Ministry of Atomic Energy cannot stoop to the level of an ordinary mortal. Indeed, now that the Ministries have amalgamated,[2] the "culprit", legally speaking, has ceased to exist. Hiding behind a screen of promises, they can go on building new, potentially deadly sites of the Chernobyl type in the most thickly populated areas of the country.

For the rectifiers reality is extremely harsh. The fairy tales about being registered and fully cared for by the medical institutions, about treatment, about their state of health being monitored have turned out to be lies – and lies that, in addition, have been covered up by the USSR Ministry of Health which ordered related matters to be treated as secrets. In its program "vzglad" [viewpoint] the Central television network revealed that this order had been given on April 7, 1990.

Not one rectifier in the organization where I work has been registered as a radiation victim. For more than 3 years, I – and I'm seriously ill myself – was demanding that they should register my friends, too. It was only in March 1989 that the lists were finally presented. Because of the nature of my work, I used to travel all over the Ukraine. Many of the rectifiers that I met before my illness, especially in the rural areas, just like me, had no idea about the decree indicating that they should be registered as patients, and they had not been given preventative treatment. There had been no announcement in the press or on television. Everything was kept secret, and this suited the Ministry of Health perfectly since they did not have the resources to give proper service to such a large number of people.

Recently, the official representatives of the Ministry of Health and of the Ministry of Atomic Energy have been repeating that not a single person will receive a lifetime radiation dose of over 35 rem, that is, the maximum permissible for a human being. It had to be pointed out that the method being used here is one that was used in the bad old days – when they want to hide the truth from the public, they bring in these figures which only a few people, basically only specialists, can make sense of. Until recently, radiation doses were measured in Roentgen. Now they use (the new RBE) units – rem – which is the equivalent in biological effect of radiation on the human being, but is smaller than the Roentgen. So, if the maximum dose is 35 rem, then those 18- to 20-year-old soldier-rectifiers, and the older people called up later, who were sent home after they had received doses of 50 Roentgens had, in fact, received more than the permitted dose which was fixed later. Moreover, they say nothing about the effect of the size of a dose received all at once, about the effect of radiation on people of different ages, or about the resistance peculiarities of the immune system. No official spokesman ever says anything about what is wrong with the health of these people. And is it possible to determine, in fact, the dose received by each individual rectifier?

---

[2] The responsibilities and organizations involved have been reorganized – or at least renamed.

In May 1986, when we were working by the power station building at Chernobyl, the radiation level was different on each square meter of ground – it varied from hundreds of milli-Roentgens per hour to several Roentgens per hour (mR/h), and the readings shown by some of the personal dosimeters, which had laid in storage for decades without being checked, frequently changed several times a day. The radiation situation for many of the later rectifiers was no better. You just have to remember the rectifiers – the "biorobots" – who cleaned up the station roof by hand. What doses they received nobody knows. All this makes the treatment of the rectifiers more difficult. It means there must be an individual approach to each particular case and a carefully worked-out course of treatment.

The All-Union Radiation Medicine Research Center (AURMRC [3]) was established in Kiev to provide expert care for all who had been subjected to unshielded radiation. It was supplied with the very latest Soviet and foreign diagnostic equipment and can call upon the expertise of the best Kiev clinics. However, most of the rectifiers have no inkling that such a center exists, for only recently have fleeting references to it begun to appear.

If in the early stages of his illness the rectifier feels that something is wrong, so that he goes with this "minor" complaint to see the doctor, then he is usually regarded as a selfish, pushy sort of person trying to obtain some sort of special privileges. Yet, it is quite clear to all that selfish, pushy people were not the kind to end up exposed to unshielded radiation in the accident zone. Even when the illness starts to develop and the doctor recognizes that the rectifier is seriously ill, in order to get into the AURMRC the rectifier has to surmount the protective barrier of the regional and provincial medical institutions and then get through the center's special outpatient department. Only a few get through. After that, it remains to wait in the queue for a bed. This can take from several days to a month – and the number of patients, especially rectifiers, is growing all the time.

At the end of October 1988, the AURMRC was visited by a delegation from the Czech Academy of Sciences, headed by Academician B. E. Paton. By sheer chance, I happened to spend some time in close proximity to this delegation (I was there when treatment procedures were demonstrated). Telling the visitors about the Center, the AURMRC representative said that statistically speaking, out of each 1000 people, the rectifiers suffer three times the amount of illness experienced by nuclear power station workers, and 2.5 times more than the inhabitants of the town of Pripyat.

The AURMRC is the only institution in the Ukraine which determines whether or not illnesses are related to radiation exposure. The people working at the AURMRC, however, do their best to avoid making such a link. They are not at all interested to know whether the person consulting them suffered from

---

[3] In Russian: Vsesoyuznyi nauchno-issledovatelskii tsentr radiatsionnoi meditsiny (VNITsRM).

this or that illness before Chernobyl or not. Anyone who managed to get into the AURMRC receives a tactful warning at the start that he should not assume that his illness will be confirmed as being radiation-related, even if he has arrived with such a diagnosis from his own local medical services. The diagnosis in such a case will be changed. "In two years of examinations", admitted one of the doctors at the Center, "such a connection was confirmed in two cases, and in those cases radiation burns were involved". So the vast majority of rectifiers are discharged from AURMRC with a diagnosis of "ordinary illness", even if practically all the bodily organs are affected.

The doctors and their director – the former Ukrainian Health Minister Romanenko, who is also the Chairman of the arbitration commission dealing with appeals – do not know, or prefer not to know that such a diagnosis is a death sentence. From that moment on the AURMRC takes no further moral, or even statistical responsibility. The statistics will therefore work out as required.

After such a diagnosis the patient has practically no chance of ever getting back into the AURMRC – he will never get past the protective barrier put up by the authorities in his own area, where, as a rule, he will now face renewed accusations of radiophobia. It will be difficult enough just to get into hospital. The regional hospitals cannot offer proper treatment to radiation victims. The doctors usually treat whichever organ seems most affected, although pathological changes are evident throughout the organism. Besides, the treatment is often of a purely symbolic nature, given the lack of drugs, drips, etc.. On the black market, though, if you manage to find your way to the right place, these medicines cost their own weight in gold. After 4–6 months the case goes before the Doctors' Panel. The patient is declared to be an ordinary invalid and given a starvation pension. After that, better food which is richer in vitamins, and drugs that are hard to obtain and which cannot be bought in the normal way are equally out of the question. One's heart grieves for the rectifiers, for they are forgotten by all. What else can we expect, though? An individual life is nothing compared with the life of the whole country – and the rectifiers are finished anyway, aren't they?

Let me say a few words about myself. Until 1986, I never went near doctors. I even carried on working when I had respiratory infections. May 13 – 23, 1986, I worked at Chernobyl, just 600 meters from the wrecked Block 4. In February 1987, the retina of my right eye became detached. Twice, in March and July, I underwent operations at the Filatov Eye Hospital in Odessa in which they used lasers to reattach the retina. Being a mining engineer by profession, I know that people who work with radioactive ores often have eye trouble (like cataracts, detached retinas etc.). In our camp in the woods near Chernobyl, where we lived, we saw a lot of blinded birds. However, doctors in Poltava and Odessa have declared that they have no evidence for such a link with radiation. I went through the following year without being examined. I was not put on any register.

In August 1988, the illness knocked me completely off my feet. In the hospitals in Poltava, at the AURMRC and the Oncological Institute in Kiev, the following disorders were identified:

– a significant change in blood composition;
– abnormal blood circulation in the brain;
– markedly uneven heartbeat caused by inflammation of the heart muscle;
– considerable hair loss;
– stomach and intestinal disorders;
– severe liver damage;
– badly receding gums.

After a month of tests and treatment the AURMRC gave its verdict: "Ordinary illness – no link with exposure to radiation." At the Oncological Institute they at first suggested a pre-emptive operation, but after 10 days of tests they changed their minds, I think because of the damage to my immune system, and suggested I should have treatment for my liver back home.

"We're afraid of these Chernobylites," one of the leading specialists said to my wife, "We never know what kind of post-operative complications we are going to run into when there is such liver damage." This was a clear admission of the link between my illness and my time in the Zone when I was exposed to unshielded radiation.

From August 1989, fluid began to collect in my abdomen. I have had operations every month and they have drained between 10 and 15 liters of fluid. It is a year and 3 months since my illness started. Because of weakness and frequent pains around my liver, I spend most of my time lying down. I am losing weight at a catastrophic rate. I am also losing strength and hope. How long will they last?

7.2 Not Related to Radiation

"Dear Comrades of the Chernobyl Union:

The person writing to you is Nina Pavlovna Sgorelo, the wife of one of the unfortunate men who took part in the Chernobyl rectification operation, the wife of a man condemned to death more than a year ago. The sentence was announced to me in a telephone call from the AURMRC clinic. It took me a long time to recover from the state of shock that resulted from that call – then I started feverishly looking for ways to save my husband. But wherever I went, the answer I got from the bureaucrats was the same: he will die, we have done all we can. But I never stop thinking and hoping that he could be helped if only our own medical profession would cast away its pride and ask the foreign experts for help. The explosion at Chernobyl created a tragic laboratory for experiments on people! If "Science demands sacrifices" (as A. M. Petrisyants put it) then let Medicine demand its sacrifices openly instead of keeping them secret and stamping them with the diagnosis: "Ordinary illness – not related to ionizing radiation." What a wealth of knowledge will be lost with the death of these people! No laboratory, even in the most prestigious research institute, will produce such a volume of knowledge and experience as the people who went through the fire of Chernobyl!

So, in the name of those future generations who, like ourselves, have no guarantee that there will not be another Chernobyl, let all the Chernobyl unions appeal to all the world's doctors to take these sick people, the veterans of the Chernobyl emergency, and study (with their consent) the nature of their illnesses. Perhaps they will be able to help some of them, too! In any case, medical practice will be enriched. The experience of treating these radiation victims will tell the doctors more than the whole history of work with radioactive substances.

Those who took part in the emergency rectification work are suffering from a whole range of ordinary illnesses but these illnesses occur in a virulent and acute form because of the weakening of the body's immune systems. That is, this might well be called radiation AIDS. A remedy for this kind of AIDS is needed, too – and this will require the combined efforts of doctors of many countries.

Accept my husband as a member of the "Chernobyl Union" – Ivan Ivanovich Sgorelo, born 1938. He was sent to Chernobyl to work on the concreting of Block 4. Now he is a Group 1 invalid and seriously ill.

Our son Nikolai was born in 1969. He served for two years at a nuclear test site in Semipalatinsk. I am very concerned about his health."

With respect,
Signed, Nina Pavlovna Sgorelo
December 5, 1989

From:
The Trade Union Soviet of the Ukrainian Republic
Trade Union Soviet of Poltava Province
All-Union Radiation Medicine Research Center
of the Academy of Medical Sciences of the USSR
Ulitsa Melmikova, 53
252050 Kiev-50

To:
Comrade N. P. Sgorelo

"Dear Nina Pavlovna:

We have to inform you that the Medical Research Center has concluded that there is no connection between your husband's illness and the fact that he spent a period of time at the Chernobyl Nuclear Power Station. It is therefore not possible to give a positive reply to your request that the institute where your husband worked before his illness should pay the difference between his former salary and the pension which has been apportioned to him. Such payment would accord neither with the existing law nor with the decision of the Soviet of the working collective.

The decision as to whether an allowance of high-demand food products should be granted is made by the Ministry of Health of the Ukrainian SSR on the basis of the appropriate evidence."

Signed, B.M. Sharbenko
(Secretary, Provincial Trade Union Soviet)[4]

7.3 The Staff of the Power Station and Others

The "Chernobyl Union" soon began to receive thousands of letters from people who were in the Special Zone at Chernobyl beginning in April 1986, and who took part in the rectification operation. Their lives since that time have much in common with that of Ivan I. Sgorelo – there is not enough medicine, the doctors refuse to link their illness with their work at Chernobyl, there is no social security for people who have sickened and suffered in this way.

At the end of this chapter is a list of just the Chernobyl nuclear power station staff who have died. It was pieced together by the Chernobyl Union. It shows that it is young people who are dying.

The mortality among the Chernobyl station staff in the pre-accident years was apparently no higher than two cases in 5000 people (4×10^{-4}). Taking into account the fact that those who died after the accident in 1986 were among the 1200 people doing rectification work, the probabilities of dying for them are follows:

1986 – 2.1×10^{-2}
1987 – 0.8×10^{-2}
1988 – 1.5×10^{-2}
1989 – 1.7×10^{-2} (first six months)

That is about 100 times the risk of death usual for a safe place of work (see Chapter 9 and Table 9.1).

---

[4]   This is the typical bureaucratic reply to requests of this type. The author has dozens of signed letters like this one. The response is usually "no connection with your stay in Chernobyl", "nothing unusual" and, in particular, the standard phrase "not related to radiation".

From April 9 to 28, 1990, a group of ex-rectifiers from Byelorussia went on a hunger strike in the clinic of the Institute of Radiation Medicine. Their main demand was that the government of the republic should pass a law which would give rectifiers legal status, and also create a council of experts who would deal with rectifiers' problems. Other demands were:

– the creation of a diagnosis and treatment center and a rehabilitation center;
– the granting of proper benefits to people who received large doses of radiation while doing rectification work at Chernobyl and who, as a result, have suffered from ill-health or have become invalids;
– a declaration to the effect that Chernobyl veterans and invalids should have a status equal to that of the veterans and invalids of World War II.

A letter signed by many rectifiers and by people who live in the Strict Control Zones (patients at the clinic of the Institute of Radioactive Medicine in Minsk) was sent to the first Session of the Byelorussian Supreme Soviet. The letter's first demand was that the Supreme Soviet should create a committee to handle Chernobyl disaster rectification work in Byelorussia. The second demand was for legislation to help all those affected by the Chernobyl disaster, whether rectifiers or residents of the contaminated areas.

Another important issue has arisen recently in some of the contaminated areas. In a number of areas previously thought to have been only mildly affected by contamination, it has been discovered that radionuclides are finding their way into the food chain at a very high rate, and that people are ingesting large doses of radiation by eating local food.

No attention was paid to these areas for 5 years, however, because the amount of cesium and strontium in the soil was considered to be small. No protective measures were taken – the people consumed local milk, meat, and vegetables, as well as berries and mushrooms from the woods. The territories involved are the Levchesk, Lumenetsk, and Pinsk regions of Byelorussia.

The people in these areas have accumulated 80–100 times more cesium and strontium in their bodies than people living in the Strict Control Zones, where restrictions were introduced and, at least to a certain extent, "clean" food has been brought in (clean, if one accepts the temporary norms set by the authorities; see Table 10.3.)

While the tragedy of the former rectifiers continues, letters continue arriving, asking the Chernobyl Union for information and assistance. Three typical examples are:

"A few days ago (January 14, 1990) a great grief befell us – our son Valerii, born 1965, passed away. He took part in the rectification work at Chernobyl, where he was sent in 1986 (May–June) while doing his army service in Minsk. Our son went for a check-up at the hospital in October 1989. They discovered cancer of the left kidney and offshoots in the lungs and liver. He leaves two small children (aged 2 1/2 and 6).

Could you please let me know where we should apply, and how we should go about getting a Chernobyl veteran's pension for Valerii's children?"

Respectfully yours,
Signed, Pyotr Petrovich Suvorov
January 20, 1990

"Please could you help me? My husband, Valerii Vasil'evich Pavlenko, died on January 10, 1990. He fell ill after taking part in the emergency work at Chernobyl. (He was there from the first days, loading sand and lead into helicopters). The diagnosis was *cancer of the colon.*"

Signed, A. Pavlenko
March 26, 1989

"In the very first days after the Chernobyl accident emergency orders came for my husband and he was sent to help with the rectification. Recently, he started to feel unwell. He had constant headaches and pains in the arms and legs. He suddenly started to go blind. No matter where we went, there was no treatment for him. He had been needed before – now nobody wanted him. We had to pay to get injections for him. It was money, money everywhere – although we are supposed to have a free health service.

He could not stand it, the pains were terrible. He could not take any more. He died on December 1, 1989.

I just wanted everyone to know the terrible truth about the way that the Chernobyl rectifiers die without getting any help whatever. These are the people they wrote about, calling them heroes, saying that they had saved the country and the world."

Signed, N. Magera
April 5, 1990

Fig. 7.1. A cemetery in the Zone

Table 7.1. This list (continued on the next page) contains the names of the (original) Chernobyl Power Station staff who died between 1986 and 1989. Acute radiation sickness is abbreviated by ARS (the Russian equivalent is ostraya luchevaya bolesn' – OLB). This list has been compiled by the "Chernobyl Union". (In most cases the abbreviations referring to job and department were not translated into English.)

No.	Surname, Name(s)	Job	Born in	Dept.	Diagnosis
1986					
1	Perchuk Konstantin G.	SMTs	1952	TTs	ARS
2	Brazhnik Vyacheslav S.	MPT	1957	TTs	ARS
3	Novik Alexandr V.	MOVTO	1961	TTs	ARS
4	Vershinin Yurii A.	MOVTO	1959	TTs	ARS
5	Akimov Alexandr F.	NSB	1953	Mgmt.	ARS
6	Lelechenko Alexandr G.	ENETs	1938	ETs	ARS
7	Baranov Anatolii I.	St. DEM	1953	ETs	ARS
8	Shapovalov Anatolii I.	St. DEM	1940	ETs	ARS
9	Konoval Yurii I.	St. DEM	1942	ETs	ARS
10	Lopatyuk Viktor I.	St. DEM	1960	ETs	ARS
11	Degtyarenko Viktor M.	OGTsN	1954	RTs	ARS
12	Khodemchuk Valerii I.	St. OGTsN	1951	RTs	ARS
13	Proskuryakov Viktor V.	SIM	1955	RTs	ARS
14	Kurguz Anatolii K.	OTsZ	1958	RTs	ARS
15	Perevozchenko Valerii I.	NSRTs	1947	RTs	ARS
16	Kudryavtsev Alexandr G.	SIUR	1957	RTs	ARS
17	Toptunov Leonid F.	SIUR	1960	RTs	ARS
18	Sitnikov Anatolii A.	ZGIS	1940	Mgmt.	ARS
19	Shashenok Vladimir N.	chief eng.	1951	ChPNB	ARS
20	Kuratov Alexei V.		1956	ATP	
21	Zinchenko Valentina K.	chief techn.	1943	PTO	
22	Konchakovskii Petr S.	mgr.	1940	Mgmt.	
23	Chernaya Antonina M.		1958	Mgmt.	
24	Nikolaev Boris V.		1931	ETs	
25	Kargapolov Valerii P.		1939	TsTsR	
1987					
26	Sheijgina Elena F.		1937	TsTPK	
27	Shalamov Nikolai S.		1936	KhTs	
28	Nechiporenko Ekaterina M.	riflewoman	1953	Guards	
29	Marchenko Vasilii M.		1942	ETs	
30	Popov Valerii A.		1946	ETs	
31	Artamonov Gennadii M.		1943	RTs	leucosis
32	Shcherbina Vasilii D.		1941	ATP	
33	Sergeeva Mariya M.		1943	MCCh-126	
34	Feshchenko Alexandr S.				

Table 7.1. This list (continued from the preceding page) contains the names of the (original) Chernobyl Power Station staff who died between 1986 and 1989. Acute radiation sickness is abbreviated by ARS (the Russian equivalent is ostraya luchevaya bolesn' – OLB). This list has been compiled by the "Chernobyl Union". (In most cases the abbreviations referring to job and department were not translated into English.)

No.	Surname, Name(s)	Job	Born in	Dept.	Diagnosis
1987					
35	Chernyi Ivan F.		1939	TsTAI	
36	Serebryakov Alexandr Y.		1950	TsTAI	
37	Kuksa Vladimir M.	electric fitter	1959	TsTAI	
38	Kamshilov Vladimir M.	electric fitter	1954	TsTAI	
39	Volchkova Adelaida V.	electric fitter	1938	TsTAI	
40	Ponomarev Vladimir S.	electric fitter	1947	TsTAI	esophagus rupture
41	Ershov Yurii P.		1941	TsTAI	
42	Ribenok Stanislav L.		1938	TsDM	
43	Guryanov Yurii N.		1949	TsTsR	
44	Dzhumok Evgenii Yu.	TsTSR mgr.	1953	TsTsR	car acccident
45	Vasil'kov Vladimir P.		1941	TsTsR	
46	Zadorin Mikhail M.	dosimetr.	1940	TsRB	heart disease
47	Ryazanov Anatolii P.		1951	ATP	
48	Fomenko Valerii N.	NTTs	1948	TTs-2	car accident
49	Bronza Igor' I.			TTs-2	
50	Laushkin Yurii A.	inspector	1939	GAEN	cancer
51	Dubodel Vasilii D.		1940	ATP	
52	Los' Aleksei V.		1955	ATP	heart condition
53	Buryakovets Sergei G.		1927	ATP	
54	Zhidilin Alexandr D.		1948	ETs	heart condition
1989					
55	Zyuzin Vladimir I.	shift-master	1939	TTs	heart condition
56	Soloviev Rudolf I.	Deputy Director	1938	Mgmt.	heart disease
57	Speka Viktor N.	sen. mash. op.	1947	TTs	car accident
58	Belozertsev Yurii N.	senior operator	1940	KhTs	heart condition
59	Nikishev Petr S.		1933	ATKh	
60	Budnik Ivan A.		1937	ATKh	
61	Poznyak Vladimir A.		1951	LER	heart condition
62	Kuznetsova			MSCh-126	
63	Volkov Igor P.	foreman	1957	OYaB	cancer
64	Mishin		1941	TTs	cancer
65	Kolesov Vladimir V.	head of dept.	1941	OYaB	heart condition

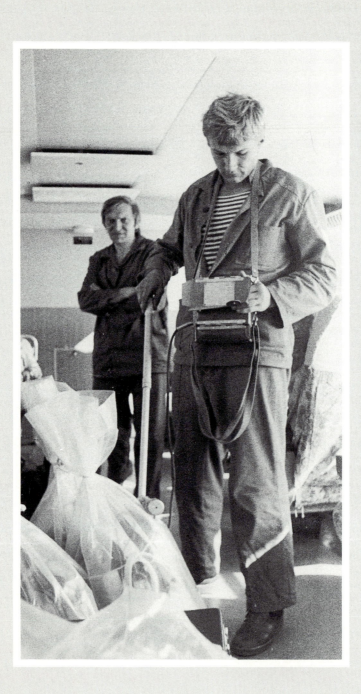

8. Radiophobia

"It is unfortunate that this state of mind should have become evident in a certain part of the population – such people live in constant expectation of damage to their health caused by radiation, not by any particular dose, just by radiation in general.

"This state of mind I have called 'radiophobia', a fear of ionizing radiation as such, in the absence of any real grounds for fear".

L.A. I'lyin
Member of the Academy of Medical Sciences of the USSR
Director of the Institute of Biophysics
Chairman of the National Radiation Protection Commission
in "Chernobyl: Events and Lessons"
(Political Literature, Moscow 1989) p. 178

Is it really fair to use "radiophobia" to describe the state in which hundreds of thousands of people find themselves after having lived 5 years in uncertainty, in the so-called "Strict Control Zones"?

From the first moments of the Chernobyl disaster the people's well being was seemingly left to fate. In the very first days the civil defense forces did not inform the populace of the threatening danger. High-ranking officials forbade raising alarm signals. Children from Pripyat and from hundreds of villages and settlements of Byelorussia, the Ukraine, and the Russian Republic, went to school on April 26, and on the following days, and took part in the May 1 celebrations. They played outside. They were outside when Block 4 was spewing millions of Curies of radiation into the atmosphere.

Is it any wonder that these "radiophobic" people think a crime was committed by those who made the negligent decision to suppress important, truthful information?

Preceding page, a technician uses a dosimeter to check the freight of a barge leaving Chernobyl.

8.1 The Head of the Regional Civil Defense

Ivan Petrovich Makarenko (M), the head of the Narodichi Regional Civil Defense since 1973, informed me (C) about the situation which arose in Narodichi Region in April 1986.[1]

C: Could you describe what happened in Narodichi Region as a result of the accident at Chernobyl in April 1986? When did you find out about the accident? How? What action was taken?

M: On April 27, as it was a Sunday, I had started doing some work on my allotment. I was planting potatoes. I started at 9 a.m. At about 11 or 11:30 I suddenly felt unwell.

I had a dryness in the mouth and my throat felt sore – a sort of roughness. I thought I was coming down with something. My allotment was down by the marshes, the floodlands of the river Uzh, towards Chernobyl. I noticed that a sort of mist was moving along the flood-lands. A greyish-brownish sort of mist. It was quite transparent. I couldn't make out what it was – although it was a sunny day. I stopped work and came home. It was just after midday. As soon as I got home the telephone rang. It was Melnik, the first secretary of the Regional Committee of the Party. He told me to get over to the Regional Committee immediately. I went there and Melnik told me that the first Secretary of Polessk Regional Committee had rung him and told him, strictly confidentially, that there had been an accident at Chernobyl. Radioactive substances had been released.

C: Who should be passing on information to Regional Civil Defense – the regional Party Secretary or the Head of Civil Defense at province or republic level?

M: It is the Civil Defense network that should be sending messages about any danger that threatens.

C: But no such message was received about the radiation danger – neither from Chernobyl, nor from provincial Civil Defense Headquarters.

M: The alarm was not raised. There was just this casual, friendly phone call from the party secretary at Polessk to our own secretary. That's how we found out. I realized immediately why I had felt ill out in the field. We sent straight away for the head of the intelligence group. He was at a textile factory. He was given the job of going straight to the stores, getting the necessary instrument, a DP-5 monitor, and bringing them to the Regional Party office. He came back with the instrument at 16:00 hours.

C: Did you have an adequate number of these instruments?

---

[1] This interview was conducted in Narodichi on March 20, 1990.

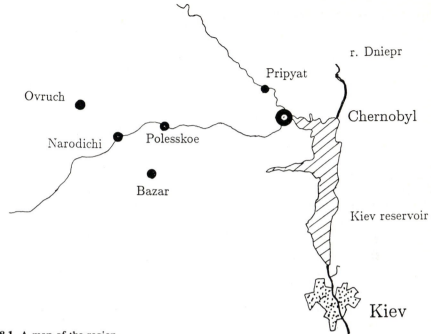

Fig. 8.1. A map of the region

M: We had just three for the whole region – one for Civil Defense use and two for the emergency medical center and the veterinary service.

C: Were they all DP-5s?

M: Yes. The health service had some old DP-2s and DP-12s, but there were no batteries for those, so we could only use the DP-5. Actually, we did not have the proper power pack for the DP-5 either and it took time to get them connected to batteries.

We took our first reading in the yard outside the regional party offices. We got a reading of 3 Roentgen/hour. What the level had been before, I have no idea – it could have been higher or lower, I don't know.

It was clear that a cloud of radioactive material had drifted along the river Uzh. That was confirmed by the drop in the level after that first reading. By 18:00 hours the level had dropped to 1.7 Roentgen/hour in the same spot.[2]

At 09:00 on April 28, it was 0.6 R/h at 13:00 it was 22 mR/h and at 18:00 it was 16 mR/h. You see how sharp the fall was. The cloud passed over and the levels dropped.

---

[2] For the remainder of this chapter, the usual scientific notation will be used for these units: "R/h" for "Roentgen/hour" and "mR/h" for "milli-Roentgen/hour" where the "milli" means "1/1000" and μR/h for "micro-Roentgen/hour" where "micro" means "1/1 000 000".

C: Did you check the soil at this time?

M: No, we didn't monitor the soil then – just the air. So I can't make any firm statement about what was on the ground.

C: Did you take any readings in the days that followed? We know that this was at the time when they started throwing bags of sand, dolomite, and lead into the reactor to stop the emissions. Did the levels go up just then?

M: No, the level never went any higher than that registered on April 27. In the village of Malyie Kleshchi, levels of 41 mR/h were recorded, then 30 mR/h. In the floodlands along the Uzh, near the village of Khistinovka, the level reached 30 mR/h. That was on May 5/6, 1986.

C: When did you start monitoring in the other settlements?

M: We started systematic monitoring about May 5/6. For example, on May 8, we got a reading of 29 mR/h in a meadow near Khistinovka. On May 9, it was down to 14 mR/h in the same spot. On May 10, it was 12 mR/h. At that time (May 8), the level in Narodichi was 0.75 mR/h. On the 9th it was 0.7 mR/h, on the 10th it was still 0.7 mR/h. It had started to stabilize.

The reactor was still releasing emissions then. On May 11, in Narodichi, we got a reading of 0.6 mR/h. On the 12th it was 0.75 mR/h, on the 13th and 14th it was 0.5 mR/h.

C: These figures are all for the same fixed locations?

M: Yes, the same reference points. The level in the meadow at Khistinovka on May 11 was 12 mR/h. On the 12th it had jumped up a bit to 14 mR/h. On the 13th it was back at 12 mR/h. On the 14th it was 11 mR/h.

The levels were fluctuating. At the village of Novoye Sharno they were: 11th of May – 11.0 mR/h; 12th of May – 13.5 mR/h; 13th of May – 14.0 mR/h; 14th of May – 10.0 mR/h.

C: Did the levels stabilize?

M: The stabilization started roughly on May 20. After that, there was a very gradual fall.

C: Are these all data from the Narodichi Civil Defense Headquarters?

M: Yes, I went out myself in the reconnaissance jeep with the scouts.

C: When did the Civil Defense get involved? After that phone call from one party secretary to another, was the alarm raised? Were the people warned of the danger?

M: There was no alarm at any time thoughout 1986.

C: None whatsoever?

M: None at all. From April 29, the province and republic Civil Defense authorities asked us for data on the radiation situation.

C: What actions did you take?

M: On April 27, at 16:30 straight after the first readings, I rang the duty officer at province Civil Defense Headquarters and reported the results.

C: You reported that you had registered 3 R/h?

M: Yes, of course, I told them that.

C: What did they have to say?

M: Nothing. They just recorded the figures. There was no order to raise the alarm. On April 28, we again reported all our readings to the commander at province Civil Defense. Just like the first time, I got no orders in reply.

C: What about help?

M: No help.

C: Did you realize that these levels put people's health in danger?

M: Of course, I realized.

C: What did you try to do?

M: When we got that reading of 3 R/h on April 27, I immediately reported it to the regional party secretary. I proposed going immediately to the control center, to sound the air-raid siren, to warn of the danger of radiation. But he forbade me to do that. He said until he got though to the Provincial Party Committee and obtained permission for it, he could not allow me to sound the alarm.

C: What did the secretary do next?

M: He went to ring up the Provincial Committee. He talked to them for a bit. Then he came back and said: 'There must be no panic and no sirens. Report to your own Civil Defense people at province level. That's all!' The regional Party Secretary is Anatolii Aleksandrovich Melnik.

C: What about the chairman of the Regional Committee?

M: He lived in the nearby village of Bazar. He did not come in on that Sunday. He turned up only on the Monday. But he couldn't do anything.

C: So what happened after that?

M: On April 28, the first refugees from the town of Pripyat started to arrive at relatives' homes in Narodichi. There was nothing organized, they just came. We found out the details of the accident from them.

We learned that there had been an accident at the nuclear power station. Radioactive emissions were still continuing – and our area was contaminated with radioactive substances.

So from the first minutes of the accident at Chernobyl people ran into a wall of secrecy. People did not get any information – or, at least, it was released in severely restricted doses. People learned about the true situation by word of mouth. And, as we all know, information passed on in that way can get distorted, by the fear, even more so as the true radiation picture was not comforting.

C: What was the situation like in Byelorussia in the first hours, days, and months after the disaster?

M: Unfortunately, the situation was very much the same as in the Ukraine. The same regime of strict secrecy stifling any mention of the disaster, of contamination levels, of dose levels, of the accumulation of radionuclides in the body. I need hardly say that in the first hours and days there was no attempt to give the children the iodine treatment that was so essential to protect their thyroid glands from radioactive iodine (despite the fact that Academician Ilyin insisted so loudly abroad that this had been done in time). Neither were the people given even basic information about the disaster.

Information was passed on by word of mouth, bit by bit. This did nothing to give people confidence. From the first minutes of the accident the authorities abandoned people to their own devices. They preferred "not to worry them".

8.2 No Panicking!

Tamara Ignatievna Grudinskaya (G) – a member of the USSR Union of Journalists and deputy editor of the Khoiniki newspaper "The Leninist Flag" told me (C) how the situation was handled in the Khoiniki region of Byelorussia.[3]

C: When did you hear about the Chernobyl accident, and what happened in your area in the first few days?

G: April 26, 1986, was a beautiful sunny day. We all were out working on the land. We wondered why so many helicopters, all painted camouflage grey, were flying in the direction of the Chernobyl nuclear power station. We decided that it was just a normal military exercise. On Monday, April 28, when I arrived at the office, my colleague Nina Konstantinovna Shabrova told me that there had, apparently, been an explosion at Chernobyl.

C: Did you believe it?

G: No. I said: "Are you crazy? Most likely this is just a dreadful rumor and a provocation. If it had really happened, we would have been told about it long before now, by radio or television."

---

[3] This interview with T.I. Grudinskaya was conducted in Khoiniki on April 15, 1991.

C: How far is it, as the crow flies, from where you were to Chernobyl?

G: 54 km. But the outlying villages, e.g., Pogonnoye, Chamkov, and Orevichi are right on the bank of the Pripyat. From there you can even see the power station.

C: And when did the first official announcement appear in the press? In your paper, for example?

G: The first short announcement about the disaster appeared in our paper only May 9, 1986.

C: And before that?

A. There was nothing. If you look through the back issues of our paper, you'll find we were describing a festive May Day procession – people full of "faith, hope", etc..

C: You say there was no information about the disaster. But during this period Block 4 was throwing out millions of curies of radioactivity and the wind was blowing all this over Byelorussia.

Were the children still outside, the older ones going to school, the little ones to the nurseries?

G: Yes – they were taking part in the May Day demonstration as well.

C: And what were the parents doing at the time?

G: Some parents acted on the rumors and the stories of the Pripyat refugees and decided to evacuate their children.

C: When?

G: From about May 6 or 7. It was after that that Dmitrii Mikhailovich Demichev, the Regional Party Secretary, warned me personally: "Don't take it into your head to move your son away from here – if you do you'll be handing over your party membership card."

8.3 The Radiation Situation in Byelorussia

Unfortunately, the same thing was happening in other places in Byelorussia, too. In Bragin, for example, the levels were up to 60 mR/h in the first few days of May and nobody said a word about the disaster at Chernobyl. No alarm was raised and no recommendations were issued with regard to the appropriate behavior in such situations.

Below, a description of the radiation situation in a number of towns and villages of Byelorussia on May 6, 1986, is given.

To get an even more precise picture, the following data should be compared to and supplemented by the ones contained in the excerpts from the report of

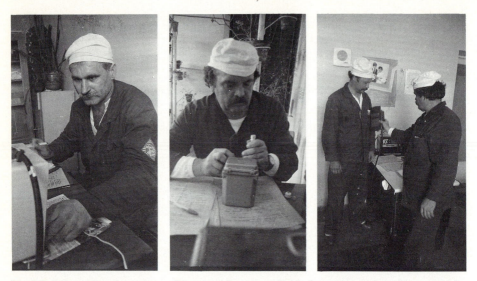

Fig. 8.2. It is good to have the traditional equipment available for measurements, however, a lot more efforts and devices are needed to be able to perform the large-scale and detailed measurements which are really required for a reliable assessment of the present situation and living conditions.

the Byelorussian Academy of Sciences reproduced in Sect. 1.6. Still further information on other regions will be found in subsequent chapters.

In the evacuation zone the gamma-background level was up to 15 000 μR/h, which is considerably higher than the level allowed for under the Basic Health Regulations-72/80 (Russian: OSP-72/80) which are 30 μR/h.

In a number of towns and villages in the southern part of Gomel province the contamination of pasture land and food was 100 or more times higher than permitted levels. For instance, the I-131 content of the soil around the village of Razin in Khoiniki region was 4 to 5×10^{-5} Ci/m^2. Around the village of Bolshiye Soroki in Narovlya region the concentration was also 4 to 5×10^{-5} Ci/m^2; in a large number of settlements in the regions of Bykhovo, Slavgorod, Cherikovo, Krasnopol'e, Kostyukovichi, and Klimovichi, the cesium-137 concentration reached 10 to 150 Ci/km^2.

Monitoring of plutonium-239 concentrations in the soil was carried out in October 1986. This showed that the borders of the territory where plutonium-139 contamination reached or exceeded 0.1 Ci/km^2 passed close to the following settlements: Yasenok, Leninskii Posyolok, Dernovichi, Teshkov (Narovlya region); Lomachi, Mokish, Bakuny (Khoiniki region); Bragin, Selets, Zarechye, Krivcha, Dubrovka, Novaya Greblya, Asarevichi, Valye, Osinnik, Golubovka, Berezki, Karlovka, Paseka, and Gden (Bragin region).

It is an unfortunate fact that even now, 5 years after the Chernobyl catastrophe, there are no detailed maps showing plutonium concentrations in the

Fig. 8.3. The map shows the location of most of the places referred to in the text

soil of a number of areas of Byelorussia. It is well known that the cesium-137 content of soil can be determined in a matter of minutes by means of a gamma spectrometer, as can other gamma-emitters. Strontium-90, however, emits only beta-radiation while plutonium-139 emits only alpha-radiation. These can be determined only by radiochemical methods which may take 10 to 15 days to produce a reading. For that reason, in the last 4 years or so, cesium-137 readings have been made at over 100 000 locations, while the strontium readings number just over 5000, and there has been even less monitoring of plutonium.

Table 8.1 shows the radiation situation in the populated areas of Bragin region as it was October 10, 1986. Short-lived elements like I-131 with a half-life of 8 days had, by this time, decayed. The gamma background level was now reduced in comparison to that in the first few days, or even months. However, there was heavy contamination with beta-particles, which is evidence of the presence of strontium-90 in the soil. Apart from this, data obtained by the Institute of Biophysics show that in the areas in question the plutonium concentration in the soil is more than 0.1 Ci/km^2. (The level of beta-contamination permitted by the RSN-76 regulations in workplaces where radioactive materials are handled is 2000 particles/$cm^2 \cdot min$.)

Table 8.1. The radiation situation in the settlements of the Bragin Rayon in Byelorussia (status: 10.10.1986). Units of β-contamination are particles/cm^2·min.

No.	Settlement	P_γ [mR/h]	caesium-137	strontium-90	β-contamination
1.	Gden'	0.13–0.21	5	1.9	9 300
2.	Karlovka	0.18–0.25	8	1.3	4 300
3.	Paseka	0.14–0.18	3	1.8	2 500
4.	Skorodnoye	0.14–2.00	3	1.1	——
5.	Lyudvinov	0.23–0.26	15	1.9	3 500
6.	Novyye Lyady	0.20–0.30	5	1.1	15 000
7.	Kamishev	0.29–0.45	5	2.3	1 900
8.	Galubovka	0.20–0.30	5	1.0	2 700
9.	Aleksandrovka	0.30–0.27	4	1.1	2 500
10.	Ivanki	0.18–0.25	5	1.2	——
11.	Osinnik	0.25–0.32	16	1.9	——

8.3.2 The Health of the People in these Regions

Let us consider the regions around Khoiniki, Borisovshchina, and Tulgovich and their particular cesium contamination. The figures given in Table 8.1 were obtained in 1987.

Table 8.2 shows the average figures for the accumulation of cesium per gram of potassium. For comparison, the table also contains the average accumulation of 92 picocuries per gram of potassium measured in America after the atomic bomb tests. After the Chernobyl accident in the Soviet Union, we began counting in tens of thousands instead of in tens. Many children are affected, and residents began to judge the contamination of their villages by the contamination of their children.

From external irradiation a human receives about 20% of his total radiation dose; the other 80% comes from contaminated food and from inhaling radioactive aerosols. These are well-known facts.

The temporarily relaxed limits on contamination in food under the maximum allowed equivalent dose VDU-88,[4] see Table 10.3, regulations allow levels of 10^{-8} Ci/liter in milk, 5×10^{-8} Ci/kg in meat. Eating food thus contaminated, a person takes in 0.7 to 0.8 rem per year in addition to external irradiation. In a lifetime, therefore, it is possible to receive a dose well above 35 rem (the guideline suggested by Academician I'lyin). However, the international norms covering such a case, where the effects of an accident last for several years, allows a dose of no more than 0.1 rem/year, i.e., a lifetime dose of 7 rem. Thus, one cannot deny that the people living in the contaminated territories have every reason to be alarmed.

---

[4] Russian abbreviation for (highest) "temporary permissible dose".

Table 8.2. For the regions of Khoiniki, Borisovshchina and Tulgovichi, activity levels (cesium) and the average figures for the accumulation of caesium per gram of potassium for the reported percentages of the population are given. For comparison the figure of the average accumulation in the United States after the atomic bomb tests is also cited.

No.	Region (activity in Ci/km^2)	Adults in [%]	Adults in [picocurie/g]	Children in [%]	Children in [picocurie/g]
1.	Khoiniki (16)	44	3000– 5000	21	4000– 5000
		20	10000–30000	30	20000–30000
2.	Borisovshchina (22)			30	30000–40000
3.	Tuglovichi (40–50)			60	35000–45000
4.	USA (average)		92		

It is wrong to dismiss their concern as that of an irrational fear of any ionizing radiation. The reader will find further confirmation of this in the chapters that follow. One way of allaying fears in the contaminated areas might possibly be a decision to stop concealing from people the truth about spending 5 years in the Strict Control Zones.

With regard to this point, it is certainly of interest to look at the (official) map of Byelorussia shown in Fig. 8.4. In the Soviet Union, picking mushrooms is extremely popular. The map indicates that it is prohibited to pick mushrooms in large areas of Byelorussia. For the unmarked regions of the republic there are no restrictions, although many say these regions should be much smaller.

8.4 People's Dosimetry

On the question of what has been called "people's dosimetry", I (C) talked again to Yevgenii Akimov (A), who, since working in the Special Zone at Chernobyl in 1986, has devoted himself to this issue.[5]

C: It is well known that very large areas suffered as a result of the Chernobyl accident and people are still worried, wondering about the dose they are absorbing now. How is the problem of radiation monitoring being tackled today?

---

[5] This discussion with Y. Akimov took place in Moscow in April, 1990.

Fig. 8.4. Map of Byelorussia. In the regions marked by larger dots it is prohibited to pick mushrooms; they are too contaminated. In the regions marked by smaller dots it is allowed to pick mushrooms, however, only under dosimetric control.

A: I am quite clear about this matter, and I want to emphasize one main thing – that 4 years after working at Chernobyl, I am still constantly finding out about affected areas that I did not know about. For example, it was only a few months ago that I found out that Kaluga province was also affected by the fallout. Quite seriously affected, too. I know, of course, that huge areas in the USSR and abroad have been contaminated – but I only found out recently that there are places in Kaluga province where people should not be allowed to live. In other words, we are again up against the same problem of lack of objective information.

I don't understand why things are handled this way, and as far as I am concerned, those who cover up the real situation are criminals. There's no other name for them! How can they let people live in areas that are ruining their health and not warn them? It's like not warning your friend when he's about to fall into a pit in the dark.

C: Do you think there is any solution? What could be done today?

A: There is no immediate, fundamental solution that I can see, however willing we are, or however many resources we pour in. The first thing, of course, would be to move people from places where they should not be living. As for that "35 rem over 70 years guideline"[6] – I am not just skeptical about it, I am totally against it.

I don't understand how they can talk like that, and these are medical experts, taking only external irradiation into account! Even if I do only absorb 35 Roentgen in my lifetime, who has counted how many I have swallowed? How many get into me from my food? How many are in the water I drink? How many are in the air I breathe? For some reason, they don't count those. And nothing is being done to set up a system whereby they could be counted. They claim that there is some program underway to supply the population with individual dosimeters, but the whole scheme is based on the wrong principle.

C: So what is your suggestion for monitoring the population? How could it be done?

A: Of all possible dosimetric methods, only one is suitable. We must have dosimeters of the integral type – that is, that give a cumulative reading. You should carry a monitor of this type wherever you go to keep a check on the dose you are building up. The dosimeters that people now have such high hopes about do not meet this requirement – they only show the level of radiation in the particular place where you happen to be. But, being unable to show a cumulative reading, they still give you no idea of your total dose – so the doctors won't know what treatment to give you. Accumulating monitors are essential.

C: But how could the system be organized, and whose job should it be?

A: As things stand at present, it is the ministerial agencies that are supposed to keep an eye on the radiation situation. But these agencies present things in a way that suits their own interests. They never get together to pool their findings and they never give a complete, detailed picture to the public. Even the diagram now available doesn't give any clear idea of the radiation exposure of the individual person, because it gives a fragmented picture. The readings are not all taken at the same time, and they are not related to actual people. The diagram gives a qualitative picture, but not a quantitative one. The readings were taken by different agencies, each working in isolation. It we want to monitor the radiation situation in any given region, or the dose absorbed by any specific person, then the information should not be collected and processed by these ministerial agencies – it should be done by other organizations which are not weighed down by all the other tasks that the agencies are overloaded with.

---

[6] Akimov refers to the suggestion of officials that a dose of 35 rem for a lifetime of 70 years is quite tolerable.

C: Could a public organization such as the Chernobyl Union provide the framework for this system?

A: I think this is not just a possibility, but a necessity. It should either be done by a restructured Chernobyl Union or some other organization should be formed to do the job. The first and most essential need is to make sure it is not a ministerial organization. It should be subordinate only to the organs of Soviet Power – those that are responsible for the lives of the people of any given region. The ministerial agencies are not accountable to the people. Only the Soviets are accountable – and they are only just becoming accountable now – so this organization must be extra-ministerial and attached to the local Soviets. All over the world there are such extra-ministerial commissions – they are even concerned with the nuclear power industry, with the production of chemicals, etc..

C: It is no secret now that hundreds of thousands of people live in the Strict Control Zones. That's the first point. Secondly, an enormous number of people live in areas where nuclear power stations are operating, or where they are being built. How long would it take to design the necessary dosimeters? And how long would it take to produce the number of dosimeters we need, once they are designed? What would such an instrument cost?

A: Well, you and I both know that dosimeters of the display type are expensive. They are beyond the means of the ordinary person. But there is another type based on thermoluminescent technology. They are cheap to produce and should therefore be cheap to buy. But we have to distinguish two categories of people who need to be equipped with dosimeters. I believe, in fact, that every person should have one. Wearing a monitor should be as normal as cleaning your teeth. Nobody is pointing out why this should be. Well, we live not only next to nuclear power stations, but also next to coal-burning and oil-burning stations – and when coal is burned practically the whole of Mendeleyev's table of elements is thrown out into the environment. That kind of fallout needs monitoring, too. That's why every person should carry an individual dosimeter – one that shows a cumulative reading.

C: Has such a meter been designed?

A: Yes, it has. I am involved in one such initial project which should lead to the issuing of such monitors to people living near atomic power stations. But to get the cost of the dosimeters down it is necessary to invest considerable sums – and I have not got the necessary resources. The ministries are not very keen to get involved. But if you asked about the price of a dosimeter on sale to the public, I would say they could cost 10 to 12 roubles each.[7]

C: Have you actually got one that works?

---

[7] In 1990, almost $ 20 (US).

A: Yes, such a meter has been made, but it is not in mass production. And the biggest obstacle in the way of mass production, as far as I can see, will be the people from the Ministry of Health and from Biophysics. Because it is not in their interests that each individual should be in possession of the facts regarding his irradiation. What would they do with these facts? The authorities realize that all kinds of questions would be raised and both answers and solutions would be required.

C: If the people no longer trust the doctors or the ministries – who have discredited themselves by keeping us in ignorance – how, then, are they going to trust the information that a "blind" instrument gives them?

A: Quite right. People certainly no longer believe what they are told by the ministries. That's why we need to create a system, based on the use of accumulating dosimeters, whereby people monitor themselves. After all, if I've got an instrument of this type, I can go to a monitoring station where I myself help to record my own reading. I actually watch the technician take the reading from my meter and I see the reading for myself. Then he gives me a document certifying my dose level. I hope the Soviets will declare this to be a legal document. The holder could then take it to a clinic and obtain the appropriate treatment on that basis. Such a document would command respect.

C: What range of measurements does this instrument give?

A: This dosimeter can measure any level from background radiation upwards. That is, if you spend a whole month in favorable conditions, subjected only to natural background radiation, then the instrument will show that. If, on the other hand, the reading is above natural background level then the question arises – where have you been? Then I can question you about where you've been, and you will help me to find the radiation source of this anomalous reading. There was a case in Kirovgrad, for example, where a radiation source was bricked up in a wall. The people who lived there were ill all the time. But if they had been carrying this sort of dosimeter and had been going for periodic check-ups, then such a thing could not have happened, could it? Should people have these instruments or not? Well, there's your answer. Even the Kirovgrad case was bad enough – what if a nuclear power station is next door?

C: If we had had dosimeters of this type in 1986, then we would not now have the problem of working out what dose millions of people have received.

A: Precisely, there would be no such problem. We would have a clear picture of any given group of people. As things are today, we cannot even be sure about the number of people who were definitely in a danger zone and who really need intensive medical care.

The same system of dosimetry would also establish a category of people who need to be kept under observation, so that any signs of deterioration in their health would be spotted. And there would probably be a third category of people whose dose is not much over natural background level. These people would not need to be kept under such constant observation. A correct diagnosis on the question "radiation related" means successful treatment.

C: So if people had had dosimeters in 1986, we would now be able to see the whole pattern.

A: Exactly. And I think we would have achieved something else as well. Our efforts could have been concentrated on the main problems instead of being dissipated as they are now. We could have spotted the most dangerous areas and concentrated on them.

As we have touched today on the subject of "people's dosimetry", I would like to say this: we need resources, we need wholehearted support. We need sponsors who could provide both money and production facilities. However, we are, unfortunately, not even in a position today to assess the extent of our needs. According to the official figure, 600 000 people worked at Chernobyl. We also know that 200 000 people should be evacuated from certain areas of the Ukraine. Another 250 000 people in Byelorussia need to be moved. We have no figures yet for Bryansk province in the Russian Federation. We need to know how many dosimeters are needed. The local Soviets should obtain this information – they are the bodies who should be concerned with these people's health. I would like to say, too, that individual dosimetry would solve yet another problem, apart from the purely medical problem. Because of the lack of information, people become anxious, and we know that anxiety can lead to illness which is not related to radiation at all, but which, nevertheless, can result in cancer. That would be another contribution that "people's dosimetry" could make. People's peace of mind is very important.

C: Shouldn't they be given dosimeters free of charge – shouldn't the State pay?

A: You are quite right to make that suggestion. I think we shall be working towards that aim. The people now living in contaminated areas should definitely be provided with dosimeters free of charge. It would be outrageous to expect them to pay for the harm that has been done to them. The sale of dosimeters at a minimum price to the rest of the population would be a different matter. But, perhaps, dosimeters should be available to all free of charge? Surely, the State could afford to provide instruments that cost 10 roubles each? For a population of 280 million that means spending about 3 billion roubles. What is 3 billion compared with the resources that we waste?

C: Who should be looking into this question of producing dosimeters?

A: The Ministry of Mechanical Engineering (which controls the nuclear power stations), the Ministry of Health, and the Soviets.

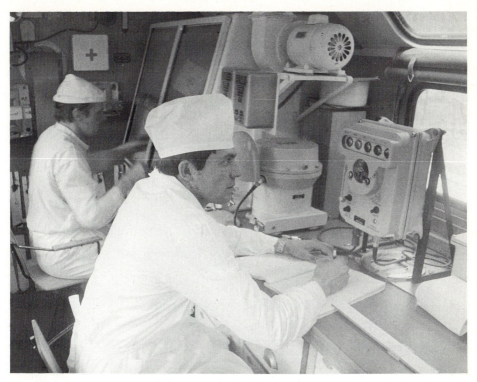

Fig. 8.5. A mobile dosimetry laboratory (Photo taken June, 1986.)

9. Hostages

"Before the Chernobyl Accident in April 1986, there had never been an accident anywhere in the world which led to heavy contamination of a large agricultural area with hundreds of thousands of inhabitants.

"Regulations designed to ensure protection from radiation in the event of an accident at a nuclear power station have been adopted all over the world. These regulations lay down guidelines as to what should be done in the initial stages of an accident to protect the population. The measures that should be taken include restrictions on time spent outdoors, iodine treatment, and evacuation. However, one thing has never been specified. No precise figure has ever been stated for the radiation dose that would be permissible over a prolonged period (e.g., in the case of people living permanently in a contaminated area). The question has never had to be considered before.

"... one can state confidently that the proposed guideline, if observed, will guarantee that there will be no measurable increase in the incidence of cancers, genetic damage or birth defects caused by irradiation of the fetus."

From the explanatory notes supplied by the USSR's National Radiation Protection Commission working group which was set up to establish a safe "lifetime dose" for people living in contaminated areas, April 1989.

Preceding page, decontamination of a truck leaving Chernobyl. The white "snow" does not signify a poor-quality photograph, but rather was caused by the film's reaction to high radiation! (Photo taken in May 1986.)

After 5 years, despite all the measures taken to ensure secrecy in all matters concerning the consequences of the Chernobyl catastrophe, including the radiation dose which the rectifiers received in the Zone and the actual radiation levels in the regions of "Strict Control", a few rare articles began appearing in the press hinting at the actual living conditions of hundreds of thousands of individuals.

Thus, the officials considered it necessary to take countermeasures. On November 22, 1988, a new document (issued by the USSR's equivalent of a Surgeon General) surfaced, which set new "lifetime radiation dose limits" for the population in the affected areas of Byelorussia, the Ukraine, and the Russian Republic. The total dose per lifetime from internal and external irradiation was set at 35 rem. This level was apparently set without measuring the natural external dose, and without a realistic monitoring of the doses already received by the residents in the 3 previous years.

To the readers, the following discussion may appear to consider only an isolated case: Chernobyl. As far as the details are concerned, that is true. However, whenever and wherever catastrophes of a similar scale occur, any government must make specific decisions based on general, often contradictory points: safety and health versus financial and technical considerations. Sometimes the situation forces the government to *temporarily* change certain tolerable and permissible norms (consider, e.g., wartime), but in no case should society tolerate that unreasonably high and even potentially dangerous "norms" are adopted as something permanent.

Using the example of "permissible" radiation doses, the following discussion also applies to other situations.

9.1 Individual Life Doses and Safety Risks

On November 22, 1988, the USSR's leading public health official, A. I. Kon-
drutsev, confirmed the individual lifetime dose limit established for people
living in the contaminated regions of the Russian Federation, Byelorussia, and
the Ukraine. These were his rulings:

1) The total individual lifetime dose of internal and external radiation should
   not exceed 35 rem.
2) The observance of the indicated limit can be verified by a check on the
   average individual equivalent dose received by members of a sample group
   in each locality.
3) The limit defined does not include the dose received from natural background radiation (in addition to the 35 rem lifetime dose, natural background radiation would give an extra dose of 24.5 rem over 30 years).

These rulings were agreed upon with the chairman of the USSR National Radiation Protection Commission (NRPC), L. A. Il'yin, on November 16, 1988.

This general lifetime limit – 35 rem over 70 years – came into force at
the same time as the 0.5 rem per year permissible dose, for a specific part
of the population, which was contained in the *"Radiation Safety Norms (RSN-
76/87)"*, also approved by the NRPC under the chairmanship of Academician
L. A. Il'yin.

What difference is there between the two documents and the two limits?

"Radiation Safety Norms 76/87" (RSN-76/86) is the basic document regulating levels of ionizing radiation. No other rules or instructions by any other
ministries or agencies are allowed to contradict it. Why then did the Ministry of
Health find it necessary to introduce a new dose limit which is in contravention
of §1.3 of the RSN?

At first glance, one might find justification in §7.2 of RSN-76/87, in the
section *"Accidental Irradiation of the Population"*: *"... depending on the scale
and character of the accident, the Ministry of Health may define temporary
basic dose limits and permissible levels."* However, it is clear that the above
limit of "35 rem over 70 years" can in no way be considered temporary. By
its definition, it is a permanent norm and, according to §1.2 of RSN-76/87, the
Ministry of Health has no right to set such a norm.

Furthermore, §7.3 (page 25) of the RSN-76/87 reads: *"Internal and external
doses received by all persons who were at any time present in a zone affected
by a nuclear accident must be assessed."*

If this is required for *"all persons ..."*, then it is required for everyone living
in the provinces and republics affected by the contamination from Chernobyl.
However, not even the Chernobyl staff or the people who took part in the

cleanup operation know their internal doses, never mind the whole population of the Russian Federation, Byelorussia, and the Ukraine. (This should even apply to those who have been checked with a radiation counter.) Therein lies another question for the Ministry of Health: Why does it not fulfill its obligations under this paragraph of the RSN?

According to the official figures the doses received by the inhabitants of the areas contaminated with radionuclides are *"below the permissible maximum"* and the contamination of foodstuffs is *"within permitted limits"*. It is not stated that the limits in question are emergency limits set up by the Ministry of Health for the 3 years following the accident, and that their period of validity expired in the summer of 1989.

Thus, it appears natural to ask the Ministry of Health: What norms are now in force?

Emergency doses received by nuclear power station workers are regulated by RSN-76/87. The maximum dose in the event of an accident is the equivalent of five times the annual permissible dose, or 25 rem. It has to be remembered that nuclear power station workers are healthy adults, who have passed strict medical checks, who work under clearly defined conditions, and who use protective clothing and equipment.

But in the stricken areas people, including women and pregnant children, live permanently in conditions exceeding an emergency situation at a nuclear power station! This has been justified by the Ministry of Health, which has set for them a maximum permissible dose of 10 rem in the first year, 3 rem in the second year and 2.5 rem in the third. Can this really be safe?

The existing norms of *Safety Risks* are based on the assumption that there is no threshold in the effect of radiation, that is, that any exposure to radiation will definitely increase the probability for the formation of malignant tumors which often cause death. This "definite result" is assessed in terms of "degree of risk". Because the human risks death in all spheres of activity and in all environments (i.e., due to earthquakes, storms, floods, etc.), the danger of irradiation can be compared to other dangers. In Table 9.1 the various risks are classified.

The starting point in the setting of radiation norms is the assertion that if 1 million people are exposed to 1 rem, 400 of them will be subjected to the risk of developing malignant tumors which will cause death.[1] In this case, the degree of risk is 4×10^{-4} per person. Since 1986 the Ministry of Health's 10 rem permissible dose automatically increased the risk to 1×10^{-3}. That is, the population's living conditions were put in the same category as jobs which are rated as "dangerous". Children, whose resistance is usually assumed

---

[1] Ye. I. Chazov, L. A. Il'yin, A. K. Guskov *"The Dangers of Nuclear War"* (A. P. P., Moscow 1982) p. 121

Table 9.1. This classification of jobs according to their safety is taken from the book by U. Ya. Margulis, 1988 *"Atomic Energy and Radiation Safety"*, page 91, Table 4.4. It should be added that risks arising from the natural environment account for 1×10^{-5} deaths per year, i.e., 10 out of every million inhabitants of the planet.

Job	Risk rating	Deaths per year	Deaths per million per year
1.	Safe	1×10^{-4}	100
2.	Relatively safe	1×10^{-4} to 1×10^{-3}	1 000
3.	Dangerous	1×10^{-3} to 1×10^{-2}	10 000
4.	Very dangerous	over 1×10^{-2}	over 10 000

to be 10 times less than an adult's, were included in the "very dangerous" category. It is not clear why the Ministry of Health permits children to live in conditions usually only allowed for healthy adults. Neither is it clear whether the Ministry, in fact, does have the right to make such a decision without the explicit agreement of the Supreme Soviet.

By setting the permissible maximum level of contamination in foodstuffs consumed in the USSR at between 1×10^{-7} and 1×10^{-8} Ci/kg, which corresponds to an annual internal radiation dose of up to 5 rem, and thus allowing more contaminated food to be sold, the Ministry of Health helped to reduce the economic losses resulting from the Chernobyl accident. Inserting the population numbers into the resulting simple expressions, the above decision and numbers seems to imply the death of an additional 200 000 people within 15 to 20 years: $4 \times 10^{-4} \times 5$ rem $\times 50$ million people eating contaminated food plus $4 \times 10^{-4} \times 10$ rem $\times 25$ million allowed to receive a dose of 10 rem.

A point to be borne in mind is that RSN-76/87 sets limits on the average figure for an individual dose in a calendar year, not for the general affected population, but for a high-risk sample. The same applies to permitted levels of contamination. What is a high-risk sample? It is *"... a small group of Category B persons (from a specified section of the population) which is homogeneous as to living conditions, age, sex, and other factors, and which is exposed to the greatest levels of radiation found within an institution or within an area under observation ..."*. In other words, our law is quite humane, requiring damage to be assessed, not on the basis of the dose received by all the people living in the observation zone (the average is 9 rem over 2 years for all the inhabitants of the zone), but only on figures obtained for that section of the people who live in the worst conditions and have received (according to a normal logarithmic distribution) significantly higher doses (up to 10 times the average). This means that in the regions where the average dose has been 9 rem there definitely must be people who have received doses of up to 90 rem. It is on the basis of this high-risk group that the law requires radiation dangers to be assessed.

9.2 What does the Dose of 35 rem per 70 Years Imply?

Neither dose levels nor cumulative dose limits are allowed to be exceeded in the case of Category B persons. So what is the meaning of 35 rem over 70 years? It is just an arbitrary set of 70 doses which a person might receive, but the size of any one dose (apart from those received in the first 3 years) is not regulated.

What can this lead to? Yu. Izrael sought to reassure the public in his article *"The Past, and a Forecast for the Future"* (*"Pravda"*, March 20, 1988):
"This norm [2] automatically ensures (given the known combination of contamination and corrective measures) a particular radiation dose (internal and external) for each year of life and concentration of foodstuffs in the basic ration. For example, a density of cesium-137 contamination up to 15 Ci/km^2 almost everywhere guaranteed the safe internal and external dose, and the permitted cesium-137 level in milk now set (at 10^{-8} Ci/l) also retains a considerable safety margin."

How *"good"* this is, is shown by a simple calculation (in which we ignore the dose received indoors, although this dose in the stricken areas is often well above background level). A contamination level of 15 Ci/km^2 corresponds to a dose of 150 micro-Roentgen/hour. Staying outdoors for two-thirds of the time gives a dose of between 0.9 and 1 rem per year. Assuming that the ratio of internal to external radiation is 1:1 (although this ratio in fact exceeds unity), we arrive at a dose of 2 rem/year in locales where there is a *"safe cesium-137 contamination level of* 15 Ci/km^2*"*. This is four times greater than the dose permitted by RSN-76/87. In just 70 years, taking into account half-life decay, the accumulated average dose will be of the order of 80 rem. This does not take into account the dose received from other radioactive isotopes, which the Ministry of Health ignores.

RSN-76/87, in Section 4, stipulates the maximum emergency dose for nuclear power station staff: *"The permitted annual dose should only be exceeded in circumstances which make this unavoidable. Each worker must be warned, must receive written communication of the fact and must give his/her consent personally to this deviation from the permitted dose. Women under 40 are not allowed to receive more than twice the permitted annual dose ... The total dose up to the age of 30 must not be more than 12 times the permitted annual dose. A single irradiation or dose increase exceeding five times the permitted annual dose should be regarded as potentially dangerous. People exposed to such levels of radiation must be released from the radiation area and sent for immediate medical examination."*

---

[2] (35 rem over 70 years)

So, it is spelled out quite clearly!

In light of these points, it is not easy to understand why the Ministry of Health considers five times the permitted annual dose to be potentially dangerous for nuclear power station staff, while 20 times (10 rem) *"does not pose any threat"* to women and children! Professor V. G. Bebeshko, the head of AURMRC, expressed this unrealistic point of view even more clearly in the newspaper *"Prapor Komunismu"* on August 1, 1989: *"... 35 rem over a lifetime is not a line dividing safe from unsafe irradiation levels; two or three times this dose will have no harmful effect on health."*

9.3 Outlook

The following lists some irregularities, discrepancies, and violations of the law, i.e., the RSN-76/87 regulations:

1) The principles on which permitted doses are based are disregarded:
   - the ratio of the permitted annual dose for nuclear power station workers and the permitted dose for the general public (excluding x-rays and background radiation) should be 10:1;
   - norms should be based on degree of risk;
   - other important factors, such as age and sex, should be taken into account.

2) In contravention of §7.2:
   - a permanent rather than a temporary limit has been set. There has been an illegal extension of the dose period from 1 year to 70 years.
   - The permitted levels are arbitrary, being unrelated to a definite dose limit.

3) A cesium-137 contamination density of 8 Ci/km^2 was fixed as the level which justified evacuation. Then, the evacuation level was raised to 15 Ci/km^2, which was finally passed as "safe" for permanent human habitation. In certain regions a level of 40 Ci/km^2 has been declared acceptable. This is a glaring infringement of the RSN restrictions of 0.06 mR/h, since at 15 Ci/km^2, this figure is 0.15 mR/h, and at 40 Ci/km^2 it is 0.4 mR/h, thus implying doses of external radiation of 1.0 R/year and 2.2 R/year.

4) The following principles of the RSN (p. 16) have been disregarded:
   - the permitted maximum dose should not be exceeded (in 1986 this was generally disregarded);
   - all unjustified irradiation should be avoided;
   - doses should be kept as low as possible.

Because these principles were ignored, presumably a large number of people received unjustifiably high doses, and 3 years after the accident a new evacuation of villages began, including some of the new settlements.

5) In contravention of §7.3 the authorities have failed to assess the internal and external irradiation of all persons in the affected zones. This makes it practically impossible to identify and evacuate people who have already received a dose of 35 rem, and those whose doses will reach this level in the future.

6) There has also been a blatant disregard of Soviet Labor Legislation since the total population, children included, has been placed in high-risk conditions without authorization from the Supreme Soviet.

The RSN states (p. 16): *"Responsibility for compliance with RSN-76/87 rests on the officials and administrators of ministries, agencies, and institutions."* Thus, legal actions should be taken to enforce the application of "normal" radiation norms for the population.

Fig. 9.1. Artifacts of the disaster (Photo taken 1990.)

Fig. 9.2. Old books and icons saved from evacuated houses in the Zone, which were frequently ransacked (Photo taken 1987.)

10. Beyond the Limit

"Blindly we build,
creating
aimlessly,
heedlessly.

"With whip, but lacking reins,
we drive
our troika on,
remorselessly.

"Small as they are,
our children
lose their hair,
their sight.

"Finally, their little hearts
give out.
Gravely, the doctors warn our wives
to bear no more.

"The grass is poisoned.
Poisoned, too, the stones,
the mushrooms in the wood,
the very dust beneath our feet."

Viktor Goncharov

The developing picture of the extent of the radiation contamination defies indifference. The first measurement devices became available and people started to learn that they are living in an area with more than 40 Ci/km^2. They learned that they are producing foodstuff containing contamination levels that are beyond exceptional dose limits. They can only be compared to those of wartime, as reflected by the officially set temporary permissible

Preceding page, monitoring and decontamination work are ongoing efforts at post-accident Chernobyl. (Photo taken in May 1986.)

levels. In some cases, 30 – 90% of the tested milk samples were higher than these norms, as well as meat, feed, and other agricultural products. Although privately owned livestock had been confiscated, animals remained on the collective farms.

In the Strict Control Zone, "clean" products were supposed to be brought in, since the home-grown produce was inedible. However, these foodstuffs were either supplied in insufficient quantities, or not at all. The 30-rouble subsidy per person and month for purchasing clean produce did not placate the public; it was nicknamed the "coffin subsidy". Consumption of contaminated locally grown foodstuffs continued to increase the radiation ingested by the people.

10.1 Worries and Emotions

Valentin Budko, first secretary of the Narodichi Regional Party Committee, talked to me about the situation in Narodichi. With a heavy heart, he told me about the tragedy of people forced to bring up little children in areas where the threat from radiation still remains.[1]

Whatever problems we try to tackle, whatever topics we discuss, everything always comes back to the situation created in this area by the accident at Chernobyl.

In the 3 years since the accident, things may have become clearer in this region, but are certainly not easier. Quite the opposite. A lot of new, unforeseen problems have arisen. Everyone here – from school children to pensioners – is worried about the effects of radiation. You can see evidence of that in the public meetings that have been held in the town of Narodichi itself, in the meetings held by staff at their places of work, in the letters sent to the higher authorities.

Following the publication in the national and local press of figures relating to radiation levels in the 30-km Zone, to permitted accumulated doses and other indicators, the people of the region have been able to compare these with the data collected here by various experts on radiology.

Many of the pronouncements of the republic's Ministry of Health and Meteorological Service [2] concerning living conditions and dose accumulation have

---

[1] The interview with V. Budko was conducted in Narodichi in February, 1990.
[2] The latter is the approximate English term for the Russian organization "Gidromet".

not been supported by the scientists of the Nuclear Research Institute of the Ukrainian Academy of Sciences or by experts in the regions and provinces.

The radiation situation remains as bad as ever. Hardly any improvement has been observed over the last 3 years. The distribution of radioactive cesium in the soil is patchy. The figures given for average dose over 70 years in the populated areas do not reflect the true state of affairs. In a large number of villages, one or two in particular, the average dose is well above 35 rem. This has been confirmed by monitoring done by visiting specialists.

The production of clean food presents very grave difficulties. Only two farms in the region – the "Gagarin" and "Dawn" collective farms – are lucky enough to have land contamination levels below 5 Ci/km^2. Even on these two farms the contamination concentration in milk is two to three times that which is permitted.

An immense amount of decontamination work has been done. This has not, however, improved the radiation level in the soil. Even in the so-called "clean" villages, 30 – 90% of the milk contamination tests have shown levels which exceed the temporary norm. Ten percent of the meat produced in this area is unsafe.

Most of the food produced in the Strict Control Zone is contaminated. Nevertheless, despite all the recommendations and restrictions this food is eaten by the local people. Medical research in the area is steadily continuing and it has come up with some very disturbing findings. It points to changes in the health patterns of the population, both of adults and children. Changes have been observed in the functioning of the thyroid glands of most of the children. The adults are suffering increasingly from a variety of disorders. There has been a rapid increase in the number of eye disorders, which has not yet been confirmed in the reports of the experts.

The deterioration in children's health is naturally causing alarm among parents and teachers. This trend is seen both in the "dirty" areas and in villages which are not in the Strict Control Zone. Experience over the last 3 years has shown that the opportunity to build up children's health in the summer months is not being fully taken advantage of. One reason for this lies in the shortness of the annual leave allowed to many women workers.

The people of the area, especially the mothers, often approach the Regional Committee of the Party with complaints, requests, and demands. In many cases we are not able to respond. For example, a certain Mrs. Grokh of Narodichi recently wrote:

"My six-year-old has swollen lymph glands. The doctors recommended moving to a different area, but my attempts to find a place to live outside this region have been unsuccessful. Please help me to move to a safe area."

We had the same request from another woman, N. V. Karas, who lives next to the regional hospital. The contamination of the soil in that area reaches 100

Ci/km^2 in places. The Regional Committee is unable to help in these cases. This causes understandable alarm and discontent, as the children's health is at stake.

With regard to the birth of mutant animals, people are asking what guarantee there is that the same thing will not be happening to humans 20 to 30 years from now. The experts on animal husbandry give no explanation for these abnormalities.

Most people in the region are extremely tense and nervous. Uncertainty is the most common state of mind, along with hopelessness. The Ministry of Health officials fail to placate. The situation is getting worse by the day and could get out of control.

A Multitude of Voices Adds to the Chorus: Anxiety, distrust, anger, and pain in the people – living for more than 5 years in contaminated territory – are understandable, in particular, in view of the irresponsible behavior discussed above. Indeed, the situation in the region is extremely serious. Out of 80 towns and villages, 69 are Strict Control Zones where people receive an additional 25% on top of their normal salary. In 41 villages people receive 30 roubles per month each in order to buy clean food. In other countries people would presumably just move out and try to make their living in uncontaminated area; even if they would have to leave a lot of their belongings behind. In the Soviet Union the situation is a bit more complicated since people have to register at the place of residence and, usually, they do not get registered if they do not have employment in the new region. In addition, there is a chronic shortage of housing in the Soviet Union which already makes moving to another residence a real problem.

To illustrate the situation, views expressed by some residents of contaminated territories are given below. They were voiced (and recorded by the author) during a meeting held in Narodichi in December 1989 to discuss the grave situation after the Chernobyl disaster.

> *Aleksandr Kaminskii, head teacher of the middle school in the village of Malyie Kleshchi:*

The situation in the village is well-known. We teachers composed a circular after we had received the results of a medical examination of our children. We discovered that in our village there is not a single child in a fully normal state of health. The children are in such a poor condition that I am amazed that they should be kept in this area. So, we composed our circular and sent it to all the parents. The response was unanimous – the parents do not want the children to start the next school year in this school. They want them evacuated immediately. I have no other name for our bureaucrats than reactionaries –

and I can see that the forces of reaction are on the offensive. We must help Gorbachev to overcome them. I see no way to do that except by revolution.

Olga Zakusilo, middle school principal:
I have taught for many years, but over the last 3 years, we have never had an assembly without one or two or three children falling to the floor. Children just drop where they stand. A tenth-class pupil fell just recently during an assembly. She hasn't got any specific illness, her legs simply give way and she falls. What causes this?

Another thing – this is supposed to be a clean village, no extra subsidies, absolutely nothing, no special deliveries of clean food. Yet the truth is that we can't drink the milk, the children can't eat any of the garden produce. And, what's more, the collective farm keeps all the cattle. We can't keep cows of our own because there's no pasture. The woods are out of bounds. Just try finding a clean place for a cow to graze. You can neither drink your own milk nor eat your own meat. Do you call that a clean village?

Grigorii Karas – secretary of the Party organization of the linen factory:
Our main worry, the main reason why we have neither peace of mind nor joy, are the circumstances brought about by the Chernobyl accident. What is the real radiation situation? How should we live and work and conduct ourselves? What will happen to our children, what will happen to us in 5 to 10 years' time? Three years have passed, but the problems are still with us.

Anatolii Tereshchuk, Chairman of the collective farm "Dawn of Communism":
I want to say something about what has happened in our region as a result of the accident of Chernobyl. I say we should send a telegram to the deputies who represent our province, telling them to raise these issues which concern us. The republic's institutions have no will to tackle these problems and the provincial leaders just keep quiet.

Pyotr Tereshchuk, machine operator of the "Petrovskii" collective farm:
It makes you feel ill just to read Loshilo's interview in "Village Life", (May 24, 1989, issue number 119). It's called, "Is it radiation that's to blame?" According to him it's all caused by mineral fertilizers, not by radiation. Why is it then that I, in fact, all our machine operators – young and healthy people – haven't got the strength to work a full shift? It wasn't like this before.

Yekaterina Yakovleva, organizer of the Malyie Minki village club:
It is clear that mistakes were made during the rectification work in 1986. People still live in villages where nobody should be living. The decision to evacuate the village has already been made, and yet they tell us that we shall be here for another 3 to 5 years.

I propose that everything said here today, from the platform and from the floor, should be recorded and sent to the Central Committee of the Party. It is possible that this appeal will reach the leaders who can make the right decision. It is clear that there are forces, quite powerful ones, that are concealing us and our predicament, afraid of being brought to account for the mistakes they have made. People like Romanenko and scientists like Likhtarev and Prisyazhnyuk,[3] among others, are involved in this. They are the ones that provide the misinformation and the doctored statistics that are printed in the papers.

Just take one figure - this maximum accumulated dose of 0.27 rem/year. I'm no radiological expert, but right next to our club there's a constant level of 0.8 milli-Roentgen/h and in the wood and in the fields it's even higher, and even I can see that people will get 10 times the dose they're supposed to be getting.

What will our children say to us later in life? For more than 3 years we have kept them here in this situation. By the autumn of this year we must evacuate all families with children.

They are doing decontamination work in our villages at the moment. They are taking off the "dirty" soil along the roadside, but where will they get "clean" soil to replace it with? The new soil will be hardly better than the soil that's removed. So why bother?

We know very well that however many bathhouses they build for us, however many sandy tracks they put asphalt on, whatever they deliver to our shops, even so, nobody who lives in the country can be completely protected from dust and nuclides. After all, we don't fly in the air, we walk on the earth, we work on the land. It's sad to see the State wasting millions on projects like these in our village and in others like it.

Galina Tretyak, pig keeper on the Soviet farm "Soviet Army":
The first thing that the Regional Committee should be doing is explaining our predicament to people so that they understand. Then they should do something

---

[3] Romanenko was the minister of health of the Ukraine. Likhtarev and Prisyazhnuk are representatives of Ukraininan Ministries.

about moving families with children out of Narodichi Region and providing homes in clean areas. That should be the Regional Committee's main aim.

> *Fyodr Silchenko, secretary of the Party organization of the collective farm "Lenin's Path":*

At our last branch meeting our members were unanimously in favor of evacuation of families with young children before the end of the year. Our school children and pre-school-age children should not come back from their summer recuperation and return to schools and nurseries in places where the radiation level is 10 times more than the permitted maximum.

10.2 Scientific Data *

According to the Meteorological Service, average cesium-137 concentrations exceed 15 Ci/km^2 in 18 populated areas. Nineteen villages have considerable areas of land where the concentration is higher than 15 Ci/km^2 and where decontamination is difficult.

The situation is made more serious by the fact that about 60 of the villages are located in close proximity to or actually within tracts of forest, where there are very high concentrations of radioactivity. For this reason, it cannot be guaranteed that the lifetime dose of the inhabitants will be less than 35 rem.

All this makes the production of clean food doubly impossible, practically across the whole region. Milk remains the most contaminated item. For example, in August 1989, 66% of the milk samples that were analyzed showed impermissible concentrations, which were in some cases 27 times greater than the norm. This contradicts the forecasts made by the experts from the Institute of Agricultural Radiobiology. At least some of its experts are encouraging the population to consume local products. It is very hard to understand the difference between these things and criminal actions.

There is no explanation, either, for the decision to declare 11 villages as "clean", although everything that they produce is just as contaminated as the produce of other localities.

The principle of basing supplementary payments to help buy clean food on the Meteorological Service's figures for soil and air contamination rather than on the contamination level of local produce is wrong.

The people's state of health is especially alarming. At the moment there are 4500 children living in the Narodichi region. According to medical examinations in 1986, all the children were classified as being in need of medical attention because of the radioactive iodine accumulated in their thyroid glands. More than 1000 people received doses of 200 rad or more in the thyroid. In

1989, eight children from this high-risk group developed third degree goiters, and these were not children from strict control areas.

Hyperplasia of the thyroid was discovered in 1986 in 437 children. There were 714 such cases in the first 6 months of 1989. Among the adult population 199 cases of hyperplasia of the thyroid were reported – 134 of these being new cases. Sixteen goiters were reported, six of these being new cases. 165 adults were found to have developed cataracts; of these, 100 were new cases. There were eight cases of cataracts in children, six of them new. Whereas in 1988, 74 new cancer cases were reported, in 1989, 74 were reported in the first 8 months. Forty-three percent of residents forced to retire on a disability pension had some form of cancer.

Further data on the alarming increase in illness is found in Chapter 1, *"Narodichi"*, and in Table 10.1 which displays figures obtained by the Republic's epidemiological unit in a selective monitoring of the doses received by a limited number of children over a period of 3 months in 1988.

Local doctors see these figures as the result of the Chernobyl accident. The officially expressed medical view is quite different. This discrepancy, together with other contradictory information from the authorities has seriously undermined the people's trust in the Ministry of Health. There is evidence indicating that doses have been understated – and not merely in isolated cases – and that the true health picture has been falsified by senior specialists in the republic's Ministry of Health. When threshold doses were calculated, doses received in the first days of the accident were not taken into account, although there are good reasons to believe that these doses were considerable.

A group of experts from the Radiobiology Council of the USSR Academy of Sciences has expressed deep concern with regard to the future population

Table 10.1. Radionuclide levels of children in Narodichi schools ("Middle School" corresponds to grades 1 – 10; "8-yr School" to grades 1 – 8) in the first quarter of 1988. The percent of children tested ist reported; this is nonstatistical data since not all children were tested. In brackets the percentages of the cases investigated are given.

No.	School	Total number of children tested	Dosage level			
			0.05-0.1	0.1-0.2	0.2-0.3	≥ 0.3
1.	Narodichi Middle School	38	20(62%)	16(35%)	2(4%)	
2.	Narodichi 8-yr School	40	21(52%)	17(43%)	2(5%)	
3.	"Solnyshko"	54	18(33%)	25(46%)	9(16.6%)	2(3%)

of the region. The Council of Ministers of the USSR and of the Ukrainian republic adopted in May/June 1989, a proposal to evacuate the population of 12 towns and villages. 1425 families (3300 people) are affected, including 319 families (1300 people) with children. A decision regarding the evacuation of two more villages is expected soon.

A definite decision had been made to evacuate families with children before September 1, 1990. However, by then only 130 families – 41% of families with children – had actually been moved out.

Because the evacuation process has been so poorly organized, 166 young families with children are still living in the villages in question. This means that 240 children (158 of school age and 82 under 7 years) are still in the village. This has forced the middle schools in the villages of Velikiye Kleshchi and Malyie Kleshchi to remain open. Sixteen children from Osioshnaya have to travel to Ragovka in Kiev province to attend school. And the kindergartens are still open.

The main reason for the delay has been the decision of the Republic's government to leave all the arrangements to the authorities in Zhitomir province, thus stretching out this vital operation over the period 1989–93.

The province is not able to provide the necessary housing without help from the Republic and the new settlements for the evacuees are still in the initial planning stage.

This is a repeat of the experience of 1986, when the resettlement of four villages, which was handled entirely at province level, stretched over more than 7 months. The evacuees have not been allowed to resettle elsewhere outside the province where they have close relatives.

Nothing has been done to solve the special problems of families caring for invalid or elderly family members.

The evacuation of the 12 villages, the problems of the Strict Control Zones, and the plight of the families with children have become political issues.

Anxiety, discontent, and mistrust over the promises and actions of the State are on the increase. Citizens are demanding action from the local authorities, although these problems can only be dealt with at a national level.

Table 10.2 contains official figures related to the radiation situation in Narodichi Region, Zhitomir province, Ukraine. They were obtained from a gamma-radiation survey taken April – May 1988, and indicate that around 40% of the region's total territory is affected.

According to the region's Radiological Research Unit, fresh animal fodder on most of the collective farms contains concentrations which exceed the temporary permitted level (VDU-88), and fluctuate between 2.2×10^{-8} and 6.8×10^{-7} Ci/kg. Milk produced on collective farms and by private sector holdings in 80 villages contains concentrations as much as 20 times greater

Table 10.2. Official results of the gamma radiation survey conducted in April and May 1988 in Narodichi Region, Zhitomir province. Out of the total area of the region of 128 400 hectares there are 56 500 hectares of agricultural land, 36 400 of arable land, 8 300 of hayfields and 11 500 of pasture land.

No.	Radiation $[\mu R/h]$		$[Ci/km^2]$		Area of soil [hectares]
	min.	max.	min.	max.	
1.		0.05		5	13 300
2.	0.05	0.15	5	15	24 200
3.	0.15	0.40	15	40	11 400
4.	0.40	0.80	40	80	3 400
5.	0.80		80		2 000
6.	Total				54 300

than that allowed by the (VDU-88) regulations. Table 10.3 gives these norms for some years after 1986; the figures were issued for "official use only". It is interesting to compare the changes since 1986; they are indicating improvements although discrepancies with the international guidelines remain.

Table 10.4 shows a forecast prepared by the USSR Academy of Medical Science Institute of Biophysics in Moscow. It indicates the doses likely to be received by the people of the 12 Strict Control villages of Narodichi Region in the period 1986 – 1990, and up to the year 2060, therefore encompassing an average lifetime of 74 years. The doses based just on cesium contamination range from 9 to 20 rem for the period 1986 – 1990 and from 50 to 110 rem for a lifetime of 74 years. Certainly, the concept of "35 rem per 70 years" of the Ministry of Health will be exceeded.

Narodichi region is, however, not the worst-affected area in the Ukraine. Several other regions, notably Polesskoye, have worse radiation levels.

I remember the summer of 1986, when we sometimes had to travel by car or helicopter from Chernobyl to the urban-type settlement of Polesskoye, where some sections of the Chernobyl administration had been moved. Even then we thought it strange that the inhabitants, including children, should be out walking in the streets of the settlement. Polesskoye is about 50 km from the power station. The convoys of trucks carrying emergency loads to Chernobyl went straight through the middle of Polesskoye, the drivers wore gauze masks over nose and mouth. Radioactive dust rose in clouds as the trucks rolled through. Meanwhile, children played peacefully in the sand.

As the fourth year since the accident drew to an end, the effect on the health of the people of Polesskoye has become evident.

Table 10.3. Temporary maximum allowed cesium-134 and -137 levels in foods and drinking water, signed by the Chief Physician of the USSR[a]

Foodstuff	Allowed levels [Ci/l, Ci/kg]			
	1986	1987	1988-90	1990-93
Drinking water	$1 \cdot 10^{-8}$	$5 \cdot 10^{-10}$		
Milk[b]	$1 \cdot 10^{-7}$	$1 \cdot 10^{-8}$	$1 \cdot 10^{-8}$	$1 \cdot 10^{-9}$
Evaporated milk	$5 \cdot 10^{-7}$		$3 \cdot 10^{-8}$	
Dry milk	$1 \cdot 10^{-7}$	$5 \cdot 10^{-8}$	$5 \cdot 10^{-8}$	$1 \cdot 10^{-9}$
Cottage cheese	$1 \cdot 10^{-7}$	$1 \cdot 10^{-8}$	$1 \cdot 10^{-8}$	$1 \cdot 10^{-9}$
Cheese	$2 \cdot 10^{-7}$	$1 \cdot 10^{-8}$	$1 \cdot 10^{-8}$	$1 \cdot 10^{-9}$
Butter	$2 \cdot 10^{-7}$	$3 \cdot 10^{-8}$		$5 \cdot 10^{-9}$
Sour cream	$1 \cdot 10^{-7}$	$1 \cdot 10^{-8}$		
Vegetable oil	$2 \cdot 10^{-7}$		$8 \cdot 10^{-8}$	
Margarine	$2 \cdot 10^{-7}$		$8 \cdot 10^{-8}$	
Meat	$1 \cdot 10^{-7}$	$5 \cdot 10^{-8}$		$1 \cdot 10^{-9}$
Poultry	$1 \cdot 10^{-7}$		$5 \cdot 10^{-8}$	
Egg[c]	$5 \cdot 10^{-8}$		$5 \cdot 10^{-8}$	
Fish	$1 \cdot 10^{-7}$		$5 \cdot 10^{-8}$	
Vegetables (roots)	$1 \cdot 10^{-7}$	$1 \cdot 10^{-8}$	$2 \cdot 10^{-8}$	$1 \cdot 10^{-9}$
Green vegetables	$1 \cdot 10^{-7}$		$2 \cdot 10^{-8}$	
Potatoes	$1 \cdot 10^{-7}$	$1 \cdot 10^{-8}$	$2 \cdot 10^{-8}$	$1 \cdot 10^{-9}$
Fresh fruit	$1 \cdot 10^{-7}$		$2 \cdot 10^{-8}$	
Dried fruit	$1 \cdot 10^{-7}$	$3 \cdot 10^{-7}$	$3 \cdot 10^{-7}$	$1 \cdot 10^{-9}$
Juices	$1 \cdot 10^{-7}$			
Grains, cereals	$1 \cdot 10^{-8}$			
Bread	$1 \cdot 10^{-8}$		$1 \cdot 10^{-8}$	
Sugar	$5 \cdot 10^{-8}$		$1 \cdot 10^{-8}$	
Mushrooms (wild)	$5 \cdot 10^{-7}$	$5 \cdot 10^{-8}$		
Mushrooms (dried)		$3 \cdot 10^{-7}$		
Baby food		$1 \cdot 10^{-8}$	$1 \cdot 10^{-9}$	
Herbs (dry)	$5 \cdot 10^{-7}$			

[a] 1986 by P.N. Burgasov (May 30, 1986, Report No. 129-252.DSP); 1987 A.I. Kondrusev (December 15, 1987, Report No. 129-252-1.DSP), 1988 and 1990-93, as in 1987 (October 6, 1988, Report No. 129-252-2.DSP)

[b] After June 1, 1986: $1 \cdot 10^{-8}$

[c] Ci/egg

One of the many letters sent to the Chernobyl Union by people living in Polesskoye will illustrate their suffering, 250 citizens of Polesskoye signed it on behalf of their families:

We, the people of Polesskoye, have been living in a Strict Control Zone for 4 years now. You probably know about the conditions in which our children live. They are not merely ill, they are being snuffed out like little candles before our eyes. Every day they complain of headaches and nausea. They are often

Table 10.4. Current measured cesium activity and radiation doses for 1986 – 1990. Projected doses until the year 2060 [in rem]. Report of the Moscow Institute of Biophysics.

Settlement	Popul.	Cesium activity			Dose 1986–90	Dose up to year 2060		
		min.	max.	av.		internal	external	total
1. Nozdrishche	485			34	11.78	12.16	34.78	58.72
2. Peremoga	65			45	13.79	14.24	22.74	50.77
3. Khriplya	58		88	32	9.61	9.92	55.78	75.31
4. Polesskoye	120			35	18.6	19.2	75.65	113.45
5. Rudnya Ososhnya	235	29	113	51	13.95	14.4	40.09	68.44
6. Slavinshchina	185			4	10.85	11.2	33.64	55.69
7. Staroye Sharno	480			31	12.09	12.48	37.48	61.84
8. Shishalovka	124	26	100	52	16.06	16.58	48.90	81.54
9. Velikiye Kleshchi	550			30	9.27	9.57	29.01	47.85
10. Malyye Kleshchi	425			34	13.64	14.08	41.82	69.54
11. Malyye Minki	179	75	84	80	20.15	20.80	60.89	101.84
12. Khristinovka	368	9	118	25	9.61	9.92	30.01	49.54

sick or have nosebleeds. They suffer from constant weakness, they go to sleep during lessons, they often faint. For 4 years this has been going on and all that time we have been waiting for the government of our country to have mercy on us and our children and let us out of this area.

Finally, in August 1989, a response came to their request that a check should be made on the radiation situation in the settlement, and that the results should be compared with the findings of the State Meteorological Service and of the Institute of Biophysics. The response came from the dosimetric monitoring unit of the then-existing "Kombinat" organization.

The aim of the proposed study was to collect the necessary data on which to base a more precise estimate of the dose that had been absorbed, to assess the scale of the necessary decontamination work, and to develop and implement radiation safety measures.

A dosimetric survey was carried out and 620 cartograms were produced of farms, private plots, and installations both inside the high-activity "patches" (where levels reached 40 Ci/km^2 and more) and elsewhere.

More than 600 environmental samples (of soil, water, air, plants and farm produce) were tested and the results analyzed. The samples were collected in the territory bounded by the averaged 15 Ci/km^2 isoline, and in certain adjacent areas. The results that were obtained show how difficult it is to establish any precise demarcation of different "patches" with their higher radiation in this territory. A concentration of cesium isotopes in the soil up to a level of 30 – 40 Ci/km^2 was, however, not unusual, and on a number of plots these figures were far exceeded. The southwest fringes of the settlement were the part of the surveyed area most heavily contaminated with cesium isotopes. On certain plots the concentration reached 120 to 300 Ci/km^2. The strontium-90 level requires further study. The maximum strontium-90 concentration discovered by radiochemical analyses was 5 Ci/km^2 found in soil samples from the leisure park area. This is twice the maximum permissible level. In samples from this same plot high plutonium readings were also obtained. The plutonium concentration was 0.084 Ci/km^2, i.e., very close to the highest permissible level for plutonium-239, which is 0.1 Ci/km^2.

There are many patches of land in and around the settlement where the cesium concentration exceeds 67 Ci/km^2, and it would be unrealistic not to expect to find higher strontium-90 concentrations on these patches. The figures obtained for concentrations of radionuclides in the air were:

cesium-137   – 1.2x10-16 Ci/liter;
strontium-90 – 2.9x10-17 Ci/liter;
all plutonium isotopes – 1.6x10-18 Ci/liter.

The alarming preliminary results obtained by the dosimetric monitoring unit of the "Pripyat" research organization show that there is an urgent need for further study. Samples of meadow plants and basic crops were selected and their cesium and strontium contents were determined. Cesium-137 concentrations in these plants varied from sample to sample, depending on the place of origin, from 2.0×10^{-7} Ci/kg to 5×10^{-10} Ci/kg.

Much higher concentrations of cesium-137 (up to 1.4×10^{-4} Ci/kg of dry weight) were found in mushrooms.

Strontium-90 levels were monitored in 29 samples and found to vary between 7.5×10^{-9} Ci/kg and 10^{-11} Ci/kg.

The USSR Ministry of Health puts Polesskoye in group 2 regarding dose levels, i.e., as one of the localities where the 35 rem lifetime dose which it uses as guideline is unlikely to be exceeded. Ministry experts consider that it might be possible to lift the emergency restrictions in such localities if a certain centrally organized package of agricultural improvement measures can be carried out and if cattle from the private sector are put on clean fodder.

However, the gamma survey showed that in 80% of the area under scrutiny the gamma level was greater than the level for Polesskoye households (0.08 micro-Roentgen/h) quoted by the experts of the Institute of Biophysics.

The situation is made even more serious by the fact that the amount of useable land is reduced by two highly contaminated zones which intersect into the settlement from the southwest and the northeast. It is not at all certain that effective decontamination of these zones will be possible.

The uncertainty surrounding the level of risk faced by people living in such difficult radio-ecological conditions casts doubt on the settlement's future. The uncertainty is compounded in the case of Polesskoye, because its inhabitants were subjected to the initial "salvo" of radiation which affected their thyroid glands, and there has been no assessment of the results of "chronic", prolonged irradiation of the whole body in the years that followed.

It is essential to make sure that the enormous amount of money spent on decontamination work, in an attempt to ensure that the 35 rem limit is not exceeded, actually secures the safety of the people of this area in the foreseeable future.

There are, however, no real grounds for confidence. The situation being what it is, it is essential to consider the option of evacuation. By now, the people of Polesskoye have lived with contamination for over 5 years. This analysis of their predicament gives clear evidence of the unreliability of the conclusions reached by the specialists of the Institute of Biophysics regarding the habitability of a number of Strict Control Zones. These conclusions are as questionable as the 35 rem per lifetime concept.

On the May 12, 1987 A. Tkachenko, First Deputy to the Chairman of the Ukrainian Council of Ministers, sent a secret "circular to all farm chiefs" and to all department heads in the State Agricultural Bureau. The circular (reference number 36-21-7-39/83) was titled, *"Supplementary Instructions concerning work to be carried out in the spring and summer of 1987 on farms located in territories subjected to radioactive contamination as a result of the accident at the Chernobyl nuclear power station."* The document dealt with routine organizational matters and asserted that the radiation situation had stabilized. It also laid down guidelines for the efficient and safe use of contaminated land. These were the instructions regarding land with radionuclide concentrations of

– up to 15 Ci/km^2:
"... work should go on as usual with selective monitoring of plant and animal products; it is recommended that dairy herds should graze on pastures where the height of the grass is not less than 10 cm";

– from 15 to 40 Ci/km^2:
"... work in the fields, sowing and planting to be carried out only after completion of the obligatory set of recommended improvement measures, unless such measures were carried out in the autumn of 1986;

> – *seek out grazing land with contamination levels no higher than 15 Ci/km^2 for pasturing of cows kept for individual use. If no such land is to be found, set aside plots for sowing of annual grasses and winter crops followed by spring crops, and use such plots for grazing or to provide fodder;*
> – *do not let milk-producing herds graze on natural pasture lands. Such lands should be given over to the fattening and recuperation of young cattle, to working animals or to dairy cows whose milk is to be used for butter. Organize semi-indoor maintenance on all collective and Soviet farms for cattle kept for milk production or whose milk is to be used for sour-milk products. These cattle are to be fed only on sown grasses*

and annual crops grown on land where contamination does not exceed 30 Ci/km². They may also be fed concentrates."

– <u>from 40 to 100 Ci/km²</u>:

"...in amendment of paragraph 3 (subheading "C") of Directive No.7-16 issued by the USSR Agricultural Bureau on August 12, 1986, prepare to carry out in 1987 a comprehensive set of compulsory improvement measures.

- *... lands with such levels of contamination are to be used for the growing of fodder crops and crops which will be sent for industrial processing (grains, rape, corn intended for silage, flax, hemp, sugar beet, etc.)*
- *Fields or individual plots where contamination exceeds 100 Ci/km² are not to be used for agricultural production. Careful radiation monitoring of such areas is to be completed by July 1 of this year. They are then to be ploughed around the perimeter with a subsoil plough. Make all necessary preparations for the exclusion of such land from further use and for subsequent forestation."*

Similar directives were also issued to agricultural managers in the contaminated areas of Byelorussia and a number of regions in the Russian Federation.

During recent years, I have often had to travel around the contaminated territories of the Ukraine and Byelorussia. I have been struck by the sight of tractor drivers wearing respirators inside their tightly shut cabs.

Between 30% and 80% of the milk, meat, and other items produced on these lands fail to meet even the requirements laid down in the regulations for temporarily permissible levels of cesium-137 and cesium-134 in food.

The first temporarily permissible levels (VDU) were issued on May 30, 1986 and were updated in December 1987 and October 1988 (see Table 10.3). Since then, there has been no attempt to introduce stricter controls. I wish to add an observation. There is no such thing as a safe concentration of radionuclides as far as living organisms are concerned. Even natural background radiation has a harmful effect on the vitality and genetic structure of human beings, animals, and plants.

We can, therefore, speak only loosely of "maximum permissible norms" and "temporarily permissible levels".

10.3 A Systematic Survey

In 1989, a systematic survey was carried out by a working group of the Scientific Research and Experimental-Production Institution of the Gosagronom of the USSR (i.e., by the deputy head A. N. Novalyaeva), of the Department of Control for the Pollution of the Environment of the Goskomgidromet of the USSR (i.e., by the chief of this Department, N. K. Gasilinoi) and of the Service for Radiation and Chemical Safety of the GO of the USSR (i.e., by its Chief, O. S. Kirillova). They studied the material related to the radiation characteristics of settlements in the Strict Control Zones of Bryansk, Kiev, Zhitomir, Gomel and Mogilev regions, as well as the data taken by an investigation of "typical" settlements. Their aim was to determine the perspectives for a normalization of the radiation situation, allowing for conditions under which the population of these areas could lead a normal life for a prolonged period of time.

Altogether, the radiation characteristics of 686 settlements were considered, out of which 222 were situated in Russia, 41 in the Ukraine, and 423 in Byelorussia.

It was stated that in 437 settlements the expected integral doses of internal and external irradiation in 70 years do not exceed 35 rem, so that life in these places could proceed without any limitations and precautions.

In the remaining settlements the normalization of the radiation situation requires soil decontamination (on the same scale of engineering efforts) combined with special agricultural improvements.

In the estimates for reducing the public exposure, the following was accepted:

- effective engineering decontamination allows to reduce the doses due to external irradiation by 1.5 times (30%);
- agricultural improvements ensure the reduction of the dose of internal irradiation up to a factor of two (the suggested measures include the yearly application of larger quantities of mineral fertilizers in the allotments of the population, and the cultivation of the natural meadows and hayfields for milk production in the private sector);
- intensive agricultural improvements ensure a reduction of the dose of internal irradiation up to four times (the suggested measures include the yearly application of larger quantities of mineral fertilizers in the allotments of the population and the creation of artificial meadows and hayfields ... with contamination levels of cesium-137 up to 40 Ci/km^2 for the private milk production of the population).

With the expected success of the suggested measures in mind, four levels of the risk population were established, as follows.

Group 1: This group contains those settlements in which a normalization of the radiation situation may be accomplished by measures reducing the doses of internal irradiation via agricultural improvements. They should result in a reduction of cesium-137 in the milk by a factor of two.

This group contains 120 settlements (see Table 10.5 and Appendix D).

Group 2: This group includes settlements in which a normalization of the radiation situation requires a combined action. Agricultural improvements are to reduce the cesium-137 content of milk up to two times; engineering work, i.e., soil decontamination of the territory of the settlement, is also necessary.

This group contains 31 settlements (see Table 10.6 and Appendix D).

Group 3: This group includes settlements in which a normalization of the radiation situation demands intensive agricultural improvements to ensure a reduction of the cesium-137 content in milk by up to four times. Simultaneously, comprehensive (engineering) decontamination of the territory has to be accomplished.

This group contains 62 settlements (see Table 10.7 and Appendix D).

Group 4: This group contains settlements in which intensive agricultural improvements together with (engineering) decontamination efforts cannot lead to a radiation level in line with the established dose limits.

This group includes 35 settlements (see Table 10.8 and Appendix D).

The investigating specialists arrived at the following conclusions:
In view of the above, it is necessary to put into practice in the first place measures leading to a normalization of the radiation characteristics as applied to groups 1, 2, and 3 of the settlements.

With regard to the settlements in group 4, special decisions are required which take into account the particulars of each of them.

Thus, the first measures towards a normalization of the radiation situation in 1989 has to be put into practice for 213 settlements. These measures include necessary engineering decontamination of the territory of 93 settlements.

Tables 10.5 to 10.8 (delegated to Appendix D) give the radiation characteristics of all the 249 settlements included in the above discussion. They give the interested reader a fair chance to study these data and to make up his own mind as to what extent he is able to agree with the conclusions offered by the above group of experts. It should be added that these lists do not include all the settlements in the contaminated areas.

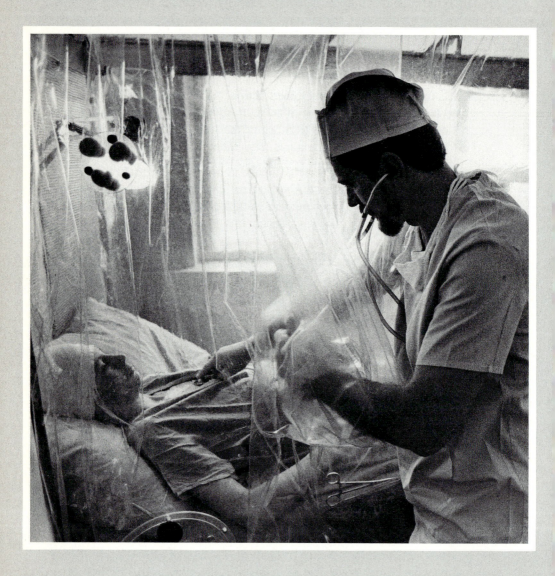

11. Doctor, Will I Live?

"We have to say very definitely that our present certainty that the Chernobyl accident did not affect the health of the population is due in no small measure to the efforts of the medical profession...

"The series of measures that we took as a nation made it possible to protect Soviet people from the possible harmful effects of radioactivity on their health. Painstaking medical research has failed to reveal any deviation from normal health patterns resulting from the effects of radiation. The health services and the special scientific units continue to monitor the health of the population in the areas affected by radioactive contamination".

Ye. I. Chazov
USSR Minister of Health, Nobel Prize Winner and Academician in "Health" (Kiev 1988) pp. 8 – 9

Objective indicators of the consequences of living in a contaminated territory and of eating "dirty", i.e., contaminated food have appeared. People have started to not just feel unwell, but to get seriously ill, and often.

The effects of living in contaminated areas are particularly obvious in children. Chronic nose and throat diseases and diseases of the gastrointestinal tract and of the liver and spleen have risen dramatically. The incidence of cataracts has increased. Birth defects have become more frequent. Enlarged thyroids among children are not uncommon.

Among adults there is increased incidence of cardiovascular complaints; swellings, diseases of cerebral blood vessels, gastrointestinal tract (gastritis, chronic cholecystitis, cholitis).

Preceding page, a radiation victim in hospital. (Photo taken at Clinic 6 in Moscow in May, 1986.)

In his *Open Letter to the General Attorney* reprinted in Chapter 1, Vasilii Yakovenko presents his account of the true situation regarding the health of the population in the affected areas of Byelorussia. We will not repeat the details here, but the reader may wish to review both the personal descriptions and hard statistical data displayed there. Yakovenko remarks that in the Khoiniki region, in 1988, there were three instances of women giving birth to terribly deformed babies – while there were 13 such cases in the first half of 1989. One of these babies had cancer of the kidneys. The number of defects per 1000 births is three to four times greater than it was in 1985, and infant mortality has risen along with it.

Another part of Chapter 1 that is of great interest and importance in this context is the section *Byelorussia* containing (part of) the results of studies conducted by the Byelorussian Academy of Sciences, a dedicated and reputable organization. Clear indications are given of deteriorating health in the contaminated areas.

Anyone listening to the speeches made by the representatives of the Health Ministries of the USSR and the Ukrainian and Byelorussian republics, or reading what they have published in the press, might well receive the impression that radiation has worked wonders for the health of the people who have been living in contaminated areas for more than 4 years. The health statistics which such officials quote in their speeches are almost always at variance with the facts as reported by the doctors who are actually treating patients in these regions. Often, one has the suspicion that the speeches are prepared, not within the Health Ministries, but by other state organs not even remotely concerned with public health.

11.1 Physicians Present Their Observations

There are countries in which, on any day, you can go to a library to look up the health statistics of a given region or of the country as a whole, and probably find current information up to the previous year. There are other countries in which such data are not just kept confidential, but are classified even as secret. According to the experience of many people, the Soviet Union does not belong to the first category. This explains why it is difficult to get reliable statistical data on public health since the Chernobyl disaster.

In this chapter, I would like to give the opportunity to medical specialists, working in Byelorussia and the Ukraine, to present their own experiences.

Fig. 11.1. A map of the Byelorussian and Ukrainian republics together with part of the Russian republic

11.1.1 Byelorussia

Tamara Byelookaya, Head of a Radiation Clinic in Minsk and one of the founders of the "Children of Chernobyl" organization, has done and is still doing much to increase public awareness of the alarming health problems now growing all over Byelorussia:[1]

Today, at last, it is clear to everyone that practically the whole population of the republic is being subjected to internal irradiation by the food it consumes. Il'yin may have said in 1986 that 30% of the population of Byelorussia had been subjected to irradiation, but now the figure is practically 100%. You only have to look at the post-mortem investigations done by our own Byelorussian scientists. They have shown that even the citizens of Vitebsk province, which is supposed to be absolutely clean, have strontium and plutonium in their bodies. The only way that these radionuclides could have got there was by being eaten.

---

[1] The author conducted this interview with T. Byelookaya in Chernobyl in May, 1990.

For over 4 years the regulations "Temporary Allowed Levels" have been in force in the republic. These regulations govern the allowed concentrations of radionuclides in foods, etc.. The last time these rules were revised was in December 1987. They should have been reviewed in April 1990, but, because the scientists could not agree about how much strontium to allow into food these regulations have never been tightened up.[2]

How long can these temporary regulations be in force? We are now into the fifth year since the disaster and our children are still drinking milk passed as fit according to the regulations authorized by A. I. Kondrusyev, the State Medical Officer of Health. That is, they drink milk which is four orders of magnitude (10 000 times) dirtier than the milk we drank in 1985. Our root vegetables are 1000 times dirtier, and so on.

If you look at the average daily intake you can see that both adults and children are absorbing 300 to 400 times more radionuclides than they did in 1985.

Because we have so few dosimeters in the republic even more heavily contaminated food reaches our tables. So the total dose absorbed by the population goes on rising and with it grows the risk of stochastic effects. For generations to come, we can be sure that the probability of genetic consequences will grow by geometric progression. We are now faced with a great number of medical problems.

Firstly, there is the question of the effect of small doses on the human organism, something that is still not fully understood. Secondly, the combined effect of the radiation and of the heavy metals which were deposited over huge areas of our republic when all kinds of different materials, primarily lead, were dropped into the burning reactor. As a result, we are now grappling with lead-induced anemia in Krasnopol and Slavgorod regions in Mogilev province. It is no better in Gomel province, either. Several regions of Brest province and also the Khoiniki region of Gomel province have been affected by boron, which causes hair loss, and lead, which can cause hypertension.

In Gomel province there has been a 14.5% increase in the number of tuberculosis cases. The specialists think that this is caused by "hot" (i.e., radioactive) particles. Experience in other parts of the world tends to suggest that radiation-related illnesses show themselves about 10 to 15 years after a radiation disaster. We are already seeing a lot of such illnesses. Endocrine pathology was the first to become evident. This is probably due to the fact that the main iodine fallout fell on areas of Polessye where goiters are endemic.

As iodine treatment had been found to be no longer necessary in these areas because some success had been achieved in an alternative treatment of goiters in the last 2 or 3 years before the accident, the thyroid glands of the people there

---

[2] See Table 11.3 for the exact numbers given in the VDU issued by the Soviet Ministry of Health, see also the discussion in Chapter 9.

were left, so to speak, defenseless when attacked by the radioactive iodine. In 1986 to 1987, thyroid monitoring was, to put it mildly, amateurish, and it is, therefore, difficult to establish what doses people actually received from the radioactive iodine in the first days after the accident.

But even if we try just to work out the collective dose, we can calculate that around 20% of the children of the southern regions of Byelorussia absorbed more than 1000 rad into the thyroid gland. According to a number of foreign researchers, even 100 rad can lead to cancer of the thyroid. A dose of 200 rad, 2 Gy, is the generally recognized threshold beyond which real damage to the thyroid is probable.

Especially in Gomel and Mogilev provinces there has been a rise in the incidence of goiters. There were 22.7 cases per 100 000 people in Gomel province in 1985. Now there are 51.9 cases per 100 000. In Mogilev province there were 21.9 cases per 100 000 in 1985. In 1989 there were 43.5 cases per 100 000 people.

Toxicosis has become 2.5 times more common in Gomel province, the number of cases having risen from 9.2 to 23.9 per 100 000. The corresponding rise in Mogilev province has taken the figure from 1.6 to 3.3.

First and second degree hyperplasia (swelling) of the thyroid gland in children is on the increase. By the end of 1989 there were some regions where 50 – 71% of the children were affected. An in-depth study of these children carried out by our clinic showed that more than 50% of them suffer from hormonal disorders. There are also a lot of auto-immune deficiencies, the so-called radiation thyroidism.

In the clinic of the Institute of Radiation Medicine, where I work, we will soon have quite a few new cases of cancer of the thyroid. There were eight such cases in 1989, and in the first 4 months of 1990 also eight.

Hypothyroidism, a reduction in thyroid activity inducing changes in the whole hormonal balance, is linked to heavy irradiation of the thyroid gland. Doses of up to 1000 rad kill the cells of the thyroid, including any cancer cells. Hypothyroidism is especially dangerous for a child as it leads to irreversible consequences and abnormal development.

The incidence of anemia in children has gone up 1.5 to 3 times since 1986. In the Chirikovo region in 1986, 13.7% of the children suffered from anemia; the figure for 1989 was 38.7%.

The number of cases of inflammation of lymph glands in the neck is rising noticeably – by up to 70% in Mogilev and Gomel provinces and by up to 20.3% in Vitebsk province.

There are also more tumors. The number of cases in Byelorussia as a whole has risen from 230 to 250 per 100 000 people over the last 3 years. In some regions, such as Krasnopolsk, the number of tumor cases has risen by 41%. In Vetkovo it has risen by 40% and in Slavgorod by 31%.

Birth defects are another growing problem in Byelorussia. There were 5.65 cases per 1000 births in 1986, and today the figure is 6.89 in the strict control zones. The figure was 5.64 cases per 1000 in Gomel province in 1986, while in 1989 it rose to 9.2 cases per 1000 births.

Infant mortality is rising in certain regions. In Khoiniki, for example, the 1986 figure was 8.8 and the 1989 figure was 14.5 per 1000 births.

We are seeing an increased number of cases of general weakness (asthenia) and all kinds of neurotic disorders. Studies show that such disorders are three to four times more common in the children of the strict zones than in the children of the relatively clean areas.

Adult health is suffering, too. Disorders affecting blood vessels in the brain are on the increase, as are stomach and intestinal disorders such as gastritis, chronic cholestitis and colitis. The whole intestinal tract is affected.

Another illness that we see more often is viral hepatitis, which is evidence of the reduced self-defense capability of the organism as a whole, and of the digestive tract in particular.

There are more and more cases of cardiovascular illnesses and illnesses affecting the nervous system of adults in the contaminated regions.

In 1990, a team was sent from our institute to the Strelichevskii village soviet in Khoiniki Region, Gomel province. They studied the local people and found that every fourth person was suffering from "vegetovascular distonia". And these were quite young people. The team noticed a rise in the number of cases of high blood pressure; it was found in 12.5% of those examined.

It should be noted, too, that the cases of high blood pressure were all serious, leading to a considerable number of cases with persistent high blood pressure and strokes. The ischemic heart disease cases were also serious.

Life in the contaminated territories is full of such psychological, social and moral stresses, creating such an instability that nervous systems and hearts cannot cope.

(*Author:* There are substantial data supplementing the above statements of T. Byelookaya, some of which is contained in Appendix D.)

11.1.2 The Ukraine

Things are not better in the Ukraine. Leonid Ishchenko, senior physician of the regional hospital of Narodichi Region, Zhitomir province, discusses the situation there:

It is now more than 3 years since the accident, but still no consensus has been reached regarding the health of the people of the region. Hundreds and hundreds of doctors have been serving here with various commissions, since 1986. Some examined the people, some checked the findings of previous commissions. An enormous amount of work has been done and it is possible, of course, to arrive at certain conclusions.

The first thing that needs to be admitted is that we were not ready with any proper contingency plan. There was no organization ready to leap into action. Even the very fact that the accident had happened was hushed up for several days. Its possible consequences were not revealed. We should have been giving emergency iodine treatment in the first hours – but we only received our iodine in this region on May 22 (delivery note Nos. 32611 and 38220.)

Then there is the question of statistics. Why is it that the 1988 reports were accepted without question, while the latest ones are being checked and double-checked? Because we want to see things as they truly are. But even now it is not possible to get a completely clear picture. When the children were examined in 1986, nine cases of lymphatic disorder were reported. The same teams of doctors were sent in again and discovered 221 cases in 1987, and none at all in 1988 and 1989. Who can believe that?

Further, obviously it would have been good to think in the first few months after the accident, or at least up to the end of 1986, about getting specialists on problems of radiation medicine to instruct the physicians of the region. An alternative could have been the organization of a special group at the Central Regional Hospital which could have been created with the cooperation of specialists (physicians and physicists) from state institutions. With such an approach a lot of problems would automatically have found their solutions. But the reality was different.

Quite a lot of scientists, professors, party workers, etc., arrived for short visits to help with clarifying explanations. They themselves ate produce which they had brought with them from the outside, they washed their hands with mineral water (brought from uncontaminated areas), without leaving their cars they measured the radiation background, they gave advice which was more anecdotal than realistic, and then they drove away. Up to this very day, the soldiers in this area walk around the territory of the region with respirators, saying that those were their orders. But for explanations and help, the people came to us, and are still coming. In a normal situation our recommendations might be met with a different response, scepticism instead of acceptance.

During all these years after the accident there has been continuous monitoring of the radiation levels due to the ingested radionuclides. Children are classified into different groups A, B, C, D, E, according to their iodine and cesium uptake. Table 11.1 shows the data.

The measurements indicate a decrease in the radionuclide uptake as the years go by. This is clearly illustrated in Table 11.2 which gives data on the cesium doses exceeding 1 microcurie. Additional numbers for people living in settlements with different degrees of contamination (regimes) are given in Table 11.3.

It is seen that there is no direct relationship between settlements with different regimes and the levels of the radiation absorbed. The doses taken up vary

Table 11.1 Classification of children according to their cumulative iodine and cesium doses

Class	Dose [in rad]		Number of children
	min.	max.	
A	0	30	1473
B	30	75	1177
C	75	200	862
D	200	500	574
E	500	≥ 500	467

Table 11.2. How doses of more than one microcurie (due to absorption of cesium) decrease with time

No.	Place	Dose [in %]		
		1987	1988	1989
1	Malyye Min'ki	77	45	10
2	Shishalovka	57	67	50
3	Narodichi	36	5	3
4	Bazar	16	7.3	7

Table 11.3. Accumulated doses (of more than one microcurie) for people living in settlements with different regimes (degrees of contamination).

No.	Place	Dose (in %)	
		1988 (7 months)	1989 (6 months)
1	Narodichi	4.8	2.9
2	Khristinovka	34.0	4.7
3	Velikiye Kleshchi	43.0	14.7
4	Rossokhovsk	66.7	35.9
5	Vyazovka	0.0	0.0
6	Guta Ksaverovskaya	25.0	0.0
7	Kalinivka	47.9	19,6

very strongly in both contaminated settlements and in clean surroundings. This is true even for members of the same family who lead essentially the same type of lifestyle and have the same type of work. To some extent this speaks for the individuality of the degree of sensitivity and uptake capacity of radionuclides.

I would like to draw attention to another fact. Preliminary analyses indicate that the level of radionuclides accumulated is not in any direct way related to the state of people's health. That implies, although one observes for several diseases a clear increase, that not all of the affected persons accumulated large doses.

In the press and in public speeches, one often hears of the increase in ailments which will take place or has already taken place. These are effects of different types: genetic, those affecting the immune system, carcinogenetic, hematological, and others. At this point, we can report the following:[3]

1) *Growth of cancer* (year – number of cases):
>   1981 – 67
>   1982 – 62
>   1983 – 69
>   1984 – 57
>   1985 – 64
>   1986 – 73
>   1987 – 81
>   1988 – 74
>   1989 – 79 in the first 9 months.

2) With the total population constantly shrinking, we have a situation in which 159 children out of every 1000 in the "clean" villages have developed hyperplasia of the thyroid gland. In the strict control areas the figure is 251 per 1000 (according to data of visiting teams of specialists).

3) In 1989, the number of cases of anemia in children increased by a factor of 4 compared to the years before. Moreover, many of these cases were chance discoveries in the course of other examinations.

4) The number of *angina victims* needing treatment in hospital grows every year. The figures are:
>   1985 – 84
>   1986 – 106
>   1987 – 101
>   1988 – 120
>   1989 – 76 first 7 months.

5) The decline in mental health is very significant. The cause may be psychogenic stress or some other factors.

6) There has been a noticeable increase in the number of cases involving vegetovascular distonia and cardiac arrythmia (as well as nervousness, headaches, and drowsiness), especially among children and young people. Other disorders, including cataracts, are also on the increase. Further research is needed, however, in this area.

---

[3] Further data and general information on the state of the health of the population may be found in the tables given in Chapters 1 and 10, and in other sections of this chapter

Further work needs to be done, also on the effects of long-term internal irradiation by radionuclides. This is made clear by the fact that different authors still recommend different allowed levels and different precautions. They also make different predictions.

We understand the complexity of the issues being raised. We understand, too, that generalizing conclusions should not be made on the basis of small statistical samples. There is also the fact that long-term poisoning with radionuclides hardly ever has specifically identifiable effects. However, we also know that the longer it takes for the effects of the accident on people's health to become apparent (if they ever do), the smaller the likelihood that we doctors will be able to do anything about them. This is especially true of any genetic effects.

I believe that certain proposals must be put forward:

1) We should, at long last, set up an office or headquarters in the central regional hospital which would be manned constantly by specialists brought in from the radiation center. These specialists should be responsible for everything that has any bearing on the health effects of the increased exposure to radiation and should, therefore, not have any cause to question our impartiality or accuse us of juggling with evidence or tailoring facts. This office should become the coordinating center for radiation matters.

2) As soon as possible, we should clarify the dose accumulated by *each individual person* and give specific advice. The sweeping formulation "average dose", as applied to populated areas, is not satisfactory.

3) We should work out the correlation between illness, including cancer, and accumulated dose.

4) There should be an end to the restrictions on the flow of information. Such restrictions only give rise to rumors and rumors do no good for morale. We are ready and willing to make information about public health matters available to the press. The seriousness of the situation should be acknowledged.

5) People should be able to choose their place of residence.

6) Maps showing in the greatest possible detail the pattern of radiation contamination in the region should be produced and made available.

7) It is time to stop merely talking about clean food (which is the main requirement if we are to prevent the accumulation of radionuclides) and actually start delivering it to the region in adequate quantities. There have never been enough fruit and vegetables and we still do not have them, either in the shops or in the hospitals.

8) It is time, too, to do something about the problem of inadequate recuperation time both for children and adults. We must extend the present, ridiculous 15 days' annual leave. Cattle have been evacuated, calves are

kept on clean fodder for 3 months before slaughter, but people have to be content with 2 or 3 week's leave.

9) The problem of inadequate medical staffing levels has been with us for 3 years and has not been solved.

11.2 Systematic Blood Tests on Children in Polesskoye

Early in 1990, a group of 18 eminent Kiev hematologists, led by the head of a hematological clinic, carried out blood tests on the children of Polesskoye region, Kiev province:[4]

Twelve hundred children, whose ages ranged from 2 to 16 years, were tested. They were drawn from the following nurseries and schools: "Beriozka" kindergarten, village of Ragovka; "Veselka", "Solnyshko" and "Druzhba" kindergartens, village of Lugoviki; schools in the urban-type settlement of Vilcha, in the villages of Lugoviki, Novy Mir, Martynovichi, Radynka, Zhovtnevoye, Shkneva.

The purpose of the tests was to establish hemoglobin, erythrocyte, and leucocyte levels in the blood, to take blood counts and to determine the activity of a number of different enzymes: peroxidase, acidic phosphotase, and alkaline phosphotase. Autoradiographic tests were also carried out in which blood smears were applied to photosensitive emulsions.

The results do not provide evidence of pronounced anemia or leucopenia in the children who were studied. The hemoglobin level of certain individuals in the preschool age group was as low as 100 g/liter. The number of erythrocytes was also somewhat reduced (3.1 to 3.3×10^{12} per liter). These same children had an increased total number of leucocytes – as many as 15 to 20×10^9 per litre.

Blood count deviations were clearly observable in children of both the younger and older age groups. The blood smears, which were stained by the Pappenheim method, showed an increased absolute content of eosinophiles characterized by a change in the configuration of the nuclei and the appearance of rod-like eosinophiles. Among the neutrophils were degenerative forms. There was evidence of anisocytosis, pyknosis, of hyposegmentation of the nuclei of cells, i.e., of so-called Pelgeroids. In up to 20% of the children attending these nurseries or schools, Pelgerian neutrophil forms predominated and all the children had some Pelgeroids. A shift to the left with appearance of young

---

[4] The author received this material from the physicians who carried out this analysis in Kiev in April, 1990.

forms and even of solitary myelocytes was observed. Many destroyed cells were encountered (in certain children as many as 50 per 100 leucocytes). In the lymphocytes of all the children there were cells with Reed-Sterberg-type nuclei, with pyknotic nuclei, with destroyed or irregular cytoplasm, the so-called degenerative forms of lymphocytes. No lymphopenia was observed.

There was an increased number of monocytes; among them were cells with poorly defined nuclei and vacuolization of cytoplasm. The blood counts showed clusters of thrombocytes – as many as 40 to 150 in each cluster (and from three to six clusters in the field of vision). Among the thrombocytes were degenerative gigantic forms, often with pronounced vacuolization.

In the smears of a number of children, cells of mononuclear type were discovered, such as we have in the past found in the blood of men who took part in the rectification operation at Chernobyl. Apart from mononuclears, pyknotic erythrocytes were also found in the blood smears, sometimes singly and, in many cases, in large clusters.

The 1970 work of I. A. Kassirskii and G. A. Alekseyev (p. 298 of their publication) shows that the appearance of pyknotic erythrocytes is linked with "a disruption of normal hemoglobin formation leading to acute degenerative changes in the erythrocytes, characterized by increased fragility and a tendency to fragment." Such changes in the erythrocytes are observed and described in cases of hemoblastosis caused by the presence in the erythrocytes of anomalous pathological hemoglobins. The appearance of pyknotic erythrocytes in children living in a zone of high radiation is evidently linked with its effect on hemoglobin structure.

Along with these changes, we observed signs of disrupted differentiation of leucocytes, the peculiarity of these being a change in the form of the nucleus. Nuclei of Pelgerian neutrophils have normal, evenly thickened contours with a small indentation in the center, and they are "kidney-shaped". A transition is observed from these forms to two-lobed nuclei (rather an "hour-glass"shape). Very few fully abnormal nuclei with three lobes were encountered. Along with stabform neutrophils and neutrophils with segmented nuclei, neutrophils with rounded nuclei with mature chromatin structure (pyknotic) and cytoplasm were found. Apart from Pelgerian changes in the nuclei of the neutrophils there were similar changes in the nuclei of the eosinophiles, as well as hypersegmented nuclei.

The Pelgeroids discovered in the children's smears and the partial Pelgerian anomalies found in 20% of cases may be evidence of the mutagenous effect of radiation.

In the literature the Pelgerian anomaly in leucocytes is described as a form of genetic mutation. The family character of the anomalies and inheritance according to dominant type has been proved. I. A. Kassirskii indicates (on page 758 of his work) that changes in the differentiation of cells involving Pelgerian anomalies may take place as a result of ionizing radiation. The phenomenological similarity of the genetic and somatic effects of radiation at the cellular level allows to suggest that the genetic mechanism involved may be identical (I. A. Vorobtsova, Paper No.1, *All-Union Conference of Radiobiologists, Pushchino, 1989*, vol.3, p.581).

The observed qualitative changes in the nuclei of the leucocytes, as well as their pyknosis and fragmentation, are evidence of lethal damage to the cell (P. B. Tokin *"Problems of Radiation Cytology"* (Leningrad, Meditsina 1974)).

The investigation of the activity of enzymes in the leucocytes showed that the distribution of the activity of peroxidase and alkaline phosphotase in the cells is uneven. There are cells, predominantly of the neutrophil type, with negative reaction values, which probably has a restricting effect on the specific functions of these cells.

The facts support the view that the changes in the children's blood are linked with long-term exposure to radiation. This is confirmed by autoradiographical tests that revealed the presence of radionuclides in the children's blood.

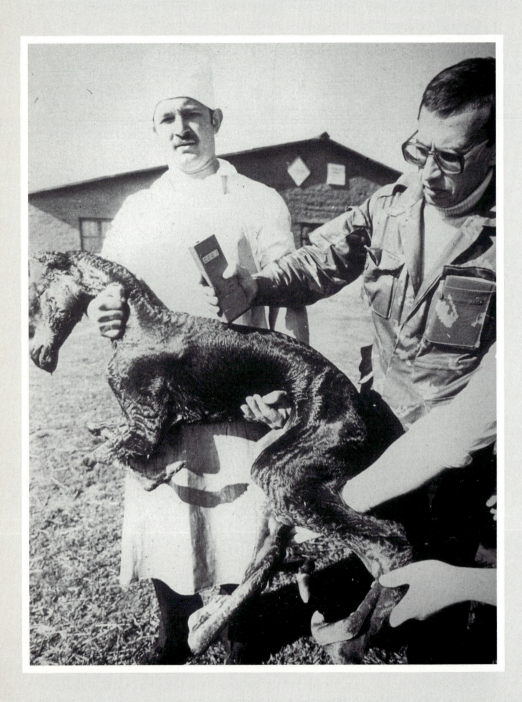

12. Mutants – What Next?

> "All the damage that
> those who fight,
> and those who hate
> do to each other
> is less
> than the damage done
> by an unskillful thought".
>> *The Dhammapada*
>> *Chapter on Thought*
>> *5th century, B.C.*

The increasing incidence of birth-defects and deformations in animals panicked the residents in the affected areas of the Ukraine and Byelorussia. I recall the anguish I felt at the birth of an eight-legged foal, which we filmed in May 1989, near Narodichi in the Zhitomir province.

The most common birth-defects in animals are the absence of one or more extremities, deformation of the skull or spine, absence of eyes, overgrowth of the eyelids, lack of hair, exposed internal organs, or absence of an anus.

The visits of various commissions have not yet resulted in any final conclusion or evaluation of these animal deformities. The question remains, "what will happen tomorrow?"

Preceding page, the author measures the radiation emitted by an eight-legged foal. (Photo taken at a farm near Narodichi, Ukraine, in May 1989.)

12.1 Effects on Plants

What is critical regarding flora and fauna is not only the density of the contamination of an area by various radionuclides, but also their form and their mobility in different kinds of soil. According to the Byelorussian Department of Agriculture, the percentage of exchange forms of radionuclides is higher on arable land.

The Radiobiology Institute of the Academy of Sciences of the Byelorussian republic has obtained some interesting data on the distribution of radionuclides in the soil.

The information comes from a study of the distribution of radionuclides in relation to the structural components of the soil: roots, the soil around the roots including the remains of the finer fibrils, and the soil containing no roots. In the swampy parts of low-lying floodlands it was shown that a quarter of the ruthenium-106 and the cesium-137 is contained in the roots, 70% is found in the soil around the roots and only 5 – 7% of the radionuclides are in the soil where there are no root systems.

In the particles of sandy soil which are free of roots, the radionuclide content is significantly higher, ranging from 26% to 59%. In the soil around the roots there may be 33% to 52%, depending on which radionuclides are involved. A higher specific activity per unit mass than in root-free soil is found in the soil around the roots. However, the specific activity of the roots is several times greater than the specific activity of the soil.

It can be concluded from this that in the northern part of the contaminated area the conditions are good for the rapid transfer of radionuclides from soil to plants.

Natural hayfields and pastures are the largest source of contaminated fodder. Here the contamination level of the vegetation depends largely on the type of soil, the moisture content of the ground, and the extent to which the land has been cultivated. Hence it is impossible to produce "clean" milk on swampy cespitose podzolized gley soils, even if the contamination level is as low as 1 Ci/km^2. "Clean" meat cannot be produced if the level is 4 Ci/km^2. The corresponding limits for wet peat soils are 4 Ci/km^2 and 12 Ci/km^2.

Drier pastures with the same types of underlying soil can produce "clean" milk where the concentration of contamination is up to 20 Ci/km^2, and "clean" meat at up to 30 Ci/km^2. On the whole, taking into account also the presence of strontium-90, it is not possible to produce clean fodder or animal products on land where cesium-137/134 contamination exceeds 15 to 20 Ci/km^2.

It should be noted that all the regulations and recommendations are based only on levels of cesium-137 and 134. The Byelorussian Agricultural Chemical Department has provided maps showing strontium-90 contamination. According to these maps the area of farmland where the strontium-90 level exceeds

0.3 Ci/km^2 is 77 000 hectares in Mogilev province, and 386 000 hectares in Gomel province.

Contaminated fodder leads automatically to secondary contamination of milk and meat. There has been an improvement in the situation regarding the contamination of milk. In Gomel (second column) and Mogilev provinces (last column) the *percentages of dirty milk* decreased significantly from 1986 to 1988:

1986 – 86.3 % – 46 %
1987 – 29.6 % – 8 %
1988 – 17.0 % – 7 %.

The percentage of dirty milk produced within particular regions strongly contaminated with radionuclides is high: The percentage over four regions of Mogilev province (Kortiukov, Cherikov, Slavgorod, and Krasnopolsk) varies between 14% and 50%, however, in the Bragin, Vetkov, Narovlya, and Khoiniki regions of Gomel province it varies from 60% to 66%.

The figures for *contaminated meat* production in Gomel and Mogilev provinces are: 1986 – 17 500 tons; 1987 – 6900 tons; 1988 – 1500 tons.

Animals kept on contaminated land and receiving fodder in which there are high concentrations of radionuclides tend to develop disorders of various physiological systems, e.g., of the endocrine, immune, and hematogenic systems.

12.2 Malformation in Farm Animals

According to the data collected by the Byelorussian Department of Agriculture, the number of disorders in farm animals in the contaminated areas began to rise from 1986 onwards. For cattle these disorders include abnormal pregnancies, as well as birth defects in calves. The number of cows failing to calve for long periods, despite repeated inseminations has also tended to rise. This is due to the failure of ovulation, the persistence of corpus luteum (yellow body), hypofunction of ovaries, metrorrhagia, and other gynecological disorders. There has also been an increase in the number of miscarriages and in the number of cases where the placenta has failed to detach after calving.

Hematological studies of cattle have revealed a significant predisposition to leucosis.

The milk yield of animals with damaged thyroid glands was two or three times smaller than that of a control group over the first 18 months after the accident. Subsequently, yields recovered and in the second year had reached 70% – 80% of the yield expected for each breed.

A special cause of alarm for the people in the contaminated areas were the increasingly frequent births of animals with various abnormalities, that is, of mutants. The main deformities seen in piglets include: absence of eyes, absence of hind legs, gross development of different parts of the body and overgrown eyelids. For calves and foals, the most common deformities include: absence of the anal opening, absence of up to three legs, absence of ears, eyes, ribs, deformities of the skull, spine, and of legs, absence of hair, underdeveloped digestive and respiratory organs, and formation of internal organs outside their proper cavity. The number of two-headed calves increased in 1989 to 90.

Figures for abnormal calves and piglets born on farms in the Narodichi region, plus data from the Slavgorod region (reflecting roughly the same picture as observed in the Khoiniki and Vetkovo regions of Gomel province) are given in Table 12.1.

In Chapters 1, 10 and 11 further information and data of relevance may be found. An international investigation of the situtation in Byelorussia and the Ukraine would certainly be helpful in clarifying the picture, but so far the path to further progress remains blocked. Naturally, the same holds true for most of the other data mentioned in this book. I hope that this book will help to provide at least a hint fo the significant problems, help to open renewed interest and honesty in solving them. Only the truth can help us to understand the scope of the problems and to find adequate answers and solutions. In the absence of global statistics, I offer again the words of an expert whose daily work includes dealing with some of the problems.

> *Anatolii Mozhar, chief officer of the Narodichi veterinary station reports:*

Cases of abnormal animal young occur mainly on farms where background radiation along with the contamination during the lifetime of the parents has been 5 to 20 times the permitted maximum. Cattle from such farms are sent to "clean" pastures for fattening.

In 1988, 1560 head of cattle with high contamination levels were sent for fattening, while 1100 head were sent in only 4 months of the year 1989.

In the last few years the number of contaminated cattle has increased. The same thing is happening in the case of pigs. Animal products from private farms have been increasingly contaminated. The commission made up of people from various levels of several different ministries which visited the region gave no firm ruling on the question of abnormalities in animals.

Table 12.1. The number of total births and the number of malformed animals in the Narodichi region. The data for 1989 refer to the first 3 to 4 months only. Rows 11 and 19 give the total numbers for calves and piglets, respectively. Overall headcount and percentages are given in rows 20 and 21. Data from the Slavgorod region is included in row 22; in this row the number for 1989 refers to the first seven months only.)

No.	Kolkhoz	Location	1987		1988		1989	
			total	malf.	total	malf.	total	malf.
	Narodichi Region							
	Calves							
1	Im. Lenina	Narodichi	396		443	3	204	2
2	Im. XXI Parts'ezda	n. Dorogin'	632		681		30	1
3	Im. Vatutina	s. Zalec'e	653		650			
4	Im. Petrovskogo	s. Motijki	479		480	8	106	1
5	Im. Kotovskogo	s. M. Kleshchi	343		378	9	17	2
6	"Chervone Polissya"	s. Zvizdal'	403	4	477	14		
7	"Iskra"	s. Golubievichi	408		519	2		
8	"Mayak"	s. Mezhiliska	381		392	1		
9	"Shlyakh Lenina"	s. V. Kleshchi	480		412		40	1
10	Im. Polyakova	s. Loznitsa	413		383		25	2
11	Sum of Calves		4588	4	4815	37	112	9
	Piglets							
12	Im. Lenina	Narodichi	1011		951	81	100	10
13	Im. Petrovskogo	s. Motijki	174		139	30		
14	private	s. Lyubarka						3
15	private	s. M. Kleshchi						6
16	"Chervone Polissya"	s. Zvisdal'	150		304	8		
17	Im. Shchorsa	s. Bazar	215		400		120	1
18	Im. Kirova	s. Sukharevka			107		41	8
19	Sum of Piglets		1550		1901	119	261	28
20	**Total**		6138	4	6716	156	373	37
21	Percentages		100	0.07	100	2.3	100	9.9
	Slavgorod Region	**Byelorussia**						
	Total Number of Deformed Animals[a]							
22				39		84		50

[a] In 1985 there were 5 animals born with deformities, and in 1986 there were 21 such births.

12.3 Experimental and Observational Studies

As a step towards an assessment of the effects of the radio-ecological situation on living creatures, Byelorussian scientists have studied specimens of plants and wild animals within the contaminated zone and compared their physiology and behavior with those of control specimens in special experimental areas outside the Zone.

The genetic effects of radiation on the chromatin apparatus of plants were revealed in a study of the frequency of mitotic anomalies in seedlings of various types of barley rye and wheat grown at an experimental station in the Khoiniki region. It was established that different strains reacted differently to radioactive contamination. The cesium-134/137 content of rye and wheat grains grown in Khoiniki was an order of magnitude higher than that of grain grown in Minsk province, while there was little difference in the case of barley. An analysis of chromosomal aberrations year by year has shown that the frequency of mitotic anomalies in plants grown in the contaminated zone in 1987 was little different from the frequency observed in ecologically clean areas. In 1988, however, in all types of plants under observation there was a noticeable rise in the frequency of chromosomal aberration.

A study of the cytogenetic effects of the accident on wild rodents, amphibians, drosophilas, agricultural animals, etc., has also confirmed that the frequency of chromosomal aberrations is $2 - 5$ times greater than in the control groups. Moreover, in the last 3 years there has been no evidence of a reduction in the amount of genetic damage, in fact, in a number of species there is tendency for damage to increase. In others the nature of chromosome damage has changed. The frequency of double division of cells has increased, as has the average number of aberrations per cell.

Experimental animals under observation in contaminated regions of Gomel and Mogilev provinces have suffered changes in thyroid function which have been shown to be abnormal by comparison with control animals. Signs of degeneration and destruction have also been noted in the pancreas, leading to increased permeability of the blood vessels, leucocyte infiltration, changes in the vessel walls, and the abnormal growth of connective tissue. At the same time, the immune system has been significantly damaged, as is shown by a change in the number of differentiated B-lymphocytes in the bone marrow, the reduction of polyferation, the differentiation and migration of T-lymphocytes, a disequilibrium of the proportions of T-helpers and T-suppressors, the growth of B-lymphoid populations among thymocytes, etc.. Damage to the structure and function of the membranes of the lymphocytes is evident, as is damage to the peripheral hormonal action mechanisms and to the system of peroxide oxidation of lipids, along with an increased level of waste products in the blood and an accompanying drop in the activity of the antioxidants system. There are marked changes in the function of the cell genome, etc..

An assessment of the functional condition of the cardio-vascular system has shown faults in the neuro-hormonal regulation mechanisms, which can be pre-pathological. Morphological research has shown structural changes in many organs and tissues. For example, structural changes of a dystrophic nature have been found in the lymphocytes of the spleen. These affect the membranes, the mitochondria, the nuclei, the endoplasmic reticulum and other structures. It is important to note that

- these changes are more marked in young animals;
- a study of the peripheral blood does not reveal the full extent of the changes taking place in the organs and tissues of, for example, the immune system;
- a direct link has been established between the extent of the changes that have been observed and both the length of time spent in contaminated territory and the levels of contamination.

13. Poisoned Waters

"Water is life!"
Ecclesiastes

Towards the end of the fifth year of the catastrophe at Chernobyl, the ecological situation in a significant part of the Dniepr river basin remains alarming. The area of the basin spans 509 000 square kilometers, extending across the territories of Byelorussia, the Russian Republic, and the Ukraine, and it contains 44 km^3 of water. It supplies water to 50 large towns and industrial centers, about 10 000 industrial enterprises and to 53 large irrigation systems which cover an area of 1.2 million hectares. Forty million people drink water from the Dniepr, and a total of 10.5 km^3 of effluent water is poured back into the river each year. Ignoring radiation, the pollutants discharged into the Dniepr in 1988 alone included 53 000 tons of organic origin, 64 800 tons of various substances in suspension, 334 000 tons of sulphates, 336 600 tons of chlorides, 3750 tons of phosphates, 15 100 tons of nitrates, and 67 tons of phenols. Also in 1988, 20 cubic kilometers of water were taken out of the river; of these, 10.3 km^3 was for industrial use, 2 km^3 was used for drinking water, and 4.6 km^3 for irrigation. In addition to continuous chemical pollution, the river is now polluted with radioactive substances.

As a result of the Chernobyl catastrophe there were, according to official statements, 50 million Curie of radioactive material deposited into the environment. According to unofficial estimates the number is 6 500 million Curie.

Preceding page, construction work to prevent the further spread of radioactive isotopes into the aquifers. (Photo taken at Chernobyl in June, 1986.)

Particularly strong radioactive pollution has been detected in the aquifers within the 30-km Zone, including the lower part of the river Pripyat and the upper part of the Kiev reservoir.

Efforts to shield the waters from radioactive pollution such as the building of dams along small rivers in the 30-km Zone, or constructing sunken walls near the nuclear power station, or the partial deepening of the upper part of the Kiev reservoir have turned out to be rather ineffective. According to the estimates of several Byelorussian scientists the sediments of the Kiev reservoir contain more than 33 000 Curie of activity.

Besides strontium-90 and cesium-137, the presence of other isotopes has been established: ruthenium-106, antimony-125, cerium-144, europium-154. It is not justified to expect a significant improvement within the near future. A large-scale state program for the monitoring and cleaning of the aquifers of the affected regions is required.

Fig. 13.1. In the first days after the Chernobyl disaster there was not too much time to prepare special equipment. Thus, one could also see fully armoured army tanks engaged in rectification work

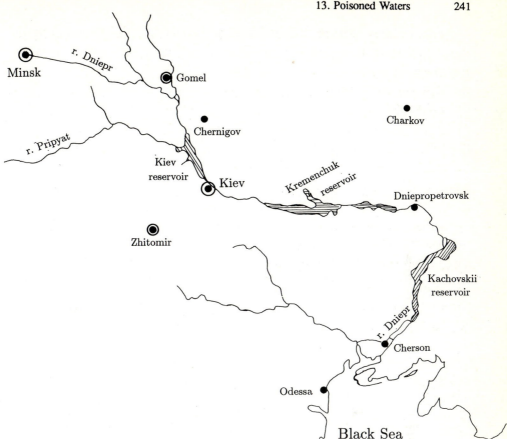

Fig. 13.2. A sketch of the main aquifiers from Byelorussia to the Black Sea

13.1 The Problem

The level of radioactive contamination in the branches of the river which flow into the Kiev reservoir reached a total beta-activity level of 10^{-7} Ci/liter in the first days after the accident, i. e., **the level was 10 000 times greater than before the accident**.

With the decay of the short-lived radionuclides the level fell back somewhat. However, the radiation situation which we now find in the reservoirs is made worse by the presence of the particularly harmful radionuclides strontium-90 and cesium-137.

It is known that radionuclides were (and still are) drained into the reservoirs from the enormous contaminated territories. This fact is reflected by the particulars of the radioactive contamination of the aquifers as a whole. Today there are high levels of contamination in the Krasnensk floodlands of the river Pripyat, of the order of 300 to 400 Ci/km^2. It is estimated by Ukrainian scientists that during the spring the land in the Krasnensk area which is covered by flood waters contributes approximately 40% to the total contamination of the Pripyat.

On these Krasnensk floodlands there is a concentration of about 9000 to 11 000 Curies of cerium-144, 6000 to 8000 Ci of cesium-137, 4200 to 4600 Ci of strontium-90, and 137 to 213 Ci of plutonium. The floodlands at Benevsk are contaminated with 400–800 Ci of cerium-144, 300–450 Ci of cesium-137, 300–400 Ci of strontium-90 and 10–25 Ci of plutonium.

According to data obtained by Ukrainian scientists, the cooling pond at the Chernobyl power station contains a concentration of 960 Ci of cerium-144, 4600 Ci of cesium-137 and 770 Ci of strontium-90.

The highest levels of contamination are to be found in the bodies of water around the Chernobyl nuclear power station. Strontium-90 levels in the waters of the Krasnensk and Semikhodovsk lake, of the Pripyat backwater, and of a number of areas of standing water on the floodlands are 1.5 to 10 times higher that the level permitted by the RSN-76/87 regulations (i.e., 5×10^{-10} Ci/liter), reaching 7×10^{-9} Ci/liter in certain places.

In the overwhelming majority of the standing and flowing surface waters the specific activity of cesium-137 fluctuates within the range 2 to 8 $\times 10^{-11}$ Ci/liter.

The Chernobyl cooling pond is an interesting exception. The specific activity of cesium-137 there varies in the range from 1 to 7 $\times 10^{-10}$ Ci/liter. The corresponding figure for the Semikhodovsk lake and the Pripyat backwater is from 2 to 4 $\times 10^{-10}$ Ci/liter.

From the time of the Chernobyl accident up through 1990, flood levels on the contaminated lands were, luckily, at no time very high (all the floods were less than 50% of the maximum expected level). Therefore, the quantity of radioactive substances washed into the Dniepr was not significant. Ukrainian scientists have estimated that in the period from 1986 to 1989, 6300 Curies of radioactive substances were carried down from the river Pripyat into the Kiev reservoir. Studies of the accumulation of radionuclides in the silt and bed deposits of the reservoirs along the Dniepr have established the figures given in Table 13.1. The calculations of the Ukrainian experts suggest that all the strontium-90 is being washed out of the bed deposits and carried down to the Black Sea.

Data from the monitoring done by various Ukrainian organizations show that the specific activity in the waters of the Pripyat and the Dniepr, as well

Table 13.1. The total activity of radioactive nuclides in the sediments of the Dniepr reservoirs [Curie]

Reservoir	Measurements of Specialists of the Ukraine Cs-137	Byelorussia Cs-137	Sr-90	Pu
Kiev	2700	4619	5617	263
Kanevskoe	500	290	40	12
Kremenchugskoe	450	2790	246	54
Dnieprodzerzhinskoe	75	65	10	10
Dnieprovskoe	35			
Kakhovskoe	200	108	7	41
Dniepro-Bugskii Liman		78	5	78
Total	3960	7950	5925	398

Table 13.2. Minimum and maximum activities of radioactive nuclei in water [Ci/l].

Reservoir, River	Cs-137 Min	Max	Sr-90 Min	Max
r. Pripyat' (Chernobyl)	$8.2 \cdot 10^{-12}$	$1.5 \cdot 10^{-11}$	$1.3 \cdot 10^{-11}$	$1.2 \cdot 10^{-10}$
r. Dniepr (Teremkovitsi)	$1.1 \cdot 10^{-11}$	$1.8 \cdot 10^{-11}$	$6.7 \cdot 10^{-12}$	$1.0 \cdot 10^{-11}$
Kiev	$3.0 \cdot 10^{-12}$	$1.5 \cdot 10^{-11}$	$2.5 \cdot 10^{-12}$	$1.4 \cdot 10^{-11}$
Kanevskaya	$5.6 \cdot 10^{-12}$	$1.5 \cdot 10^{-11}$	$6.4 \cdot 10^{-12}$	$1.4 \cdot 10^{-11}$
Kremenchugskoye (Cherkassy)	$1.3 \cdot 10^{-12}$	$1.7 \cdot 10^{-11}$	$2.9 \cdot 10^{-12}$	$8.1 \cdot 10^{-12}$
Dnieprodzerzhinskoye (Kremenchug)	$1.7 \cdot 10^{-12}$	$4.4 \cdot 10^{-12}$	$3.7 \cdot 10^{-12}$	$6.2 \cdot 10^{-12}$
Dnieprovskoye	$5.4 \cdot 10^{-13}$	$4.5 \cdot 10^{-12}$	$3.7 \cdot 10^{-12}$	$7.1 \cdot 10^{-12}$
Kakhovskoye	$4.8 \cdot 10^{-12}$	$4.0 \cdot 10^{-12}$	$1.5 \cdot 10^{-12}$	$7.5 \cdot 10^{-12}$
Dnieprovskii Liman	$8.7 \cdot 10^{-13}$	$1.8 \cdot 10^{-12}$	$6.0 \cdot 10^{-12}$	$7.3 \cdot 10^{-12}$

as those of the reservoirs along the whole course of the Dniepr is 2 to 3 orders of magnitude lower than the indicated level permitted by the RSN-76/87 regulations. Nevertheless, it should be noted that the strontium-90 and cesium-137 concentrations are 10 to 100 times higher than the pre-accident levels. Data relating to specific activity in water are given in Table 13.2.

There are times of the year, usually in summer and fall, when concentrations in the water of practically all the reservoirs exceed the limits permitted for irrigation. Since the radioactive contamination of the Ukraine is not uniform but instead is rather patchy, this factor hinders resuming farming on irrigated land. That is true especially in the southern areas of the republic, where evaporation causes accumulation (in the soil of the irrigated areas) of those radionuclides which are easily dissolved and readily taken up by plants.

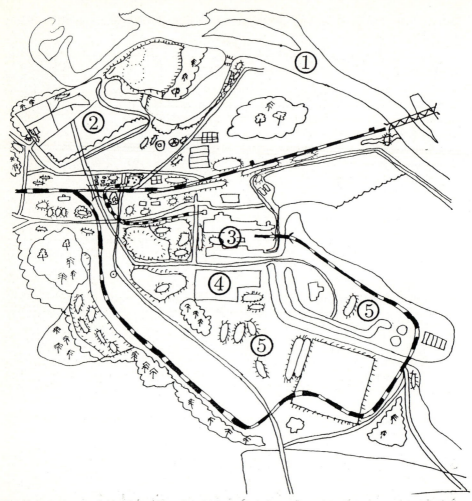

Fig. 13.3. A sketch of the 30-km Zone in which the locations of some buildings, the railway line, and the river Pripyat are indicated. The areas with "hatched" boundaries denote the emporary tombs containing altogether about 500 million tons of radioactive debris. (1 – river Pripyat, 2 – town Pripyat, 3 – reactor site, 4 – transformer station (ORU), 5 – some of the tombs; towards the left of 3 and 4 there is Phase 1 and the rectangle close to 5 indicates Phase 2)

Measurements have also been conducted on the river bed deposits of some floodlands. Here the strontium-90 content reaches 4.0×10^{-7} to 1×10^{-6} Ci/kg in situ; the concentration of cesium-137 is between 5×10^{-7} and 5×10^{-6} Ci/kg.

The most dangerous potential threats to the water resources of the Dniepr basin are, the floodlands of Krasnensk lake and Semikhodovsk lake, the Pripyat backwater, the Chernobyl cooling pond, and the temporary radioactive debris disposal dumps in the 30-km Zone (the so-called temporary "tombs").

Fig. 13.4. Heavy equipment and large scale engineering work were required to deal with the problems posed by radioactively polluted forests and soil

Fig. 13.5. Heavy equipment is needed now even more badly to deal with the radiative remains hurriedly buried in the preliminary tombs, see Fig. 13.3. Will the attempts to limit the leakage of readiation to the aquifiers, be at least partially successful? The scope of the problems is enormous

13.1.2 Biological Impact

Research done by Ukrainian scientists shows that practically all the radionuclides now in the water accumulate in living organizms which make up the food chain. Especially significant is the amount of radionuclides building up in the silt.

The "coefficients of accumulation" show how many times larger or smaller will be the quantity of radioactive contamination built up in the silt, in plant or animal organisms, in comparison to water whose assumed radionuclide content is taken to be unity. For a long time such information was kept secret and only recently did it become known that the radioactivity of water plants such as water thyme, pond weed, and cladophylls has increased 18 – 30 times.

The accumulation coefficient of cesium-137 in silt is 100 – 10 000, in crustaceans it is 10 – 20, in mollusks it is 750 – 950, in fish 45, and in cladophylls it is 62 – 1342. The accumulation coefficients of zirconium and antimony are even higher (and it should be repeated that the Chernobyl disaster released, besides strontium-90 and cesium-137, also ruthenium-103, zirconium-95, antimony-106, iodine-131, and a multitude of other intermediate decay products of uranium). They have been spread unevenly and have formed their own local radiation patches, the activity of which will, in the course of time, be carried into the water ecosystems.

Scientists have observed especially large accumulations of radionuclides in the creatures which inhabit the river bed. In terms of contamination per unit of mass the concentration in mollusks is 300 – 50 000 times greater than in flowing water. The amount of strontium in molluscs of the Kiev reservoir is 200 times greater than before the accident, while their content of radioactive cesium-134 and -137 is 500 times greater. The radioactivity of soft vegetation and animal organizms – crustaceans, mosquito larvae, etc., which form the staple diet of fish caught for human consumption is also quite high. In 1989, the average content of radioactive cesium in a selection of fish was 1.7×10^{-8} Ci/kg of live weight, while the figure for strontium-90 was 8.0×10^{10} Ci/kg.

13.2 Suggested Solutions

Water conservation measures aimed at preventing contamination of the river Pripyat have been carried out by the Ministries of Water Management of the USSR and of the Ukrainian Republic, with the assistance of experts from other agencies. In an urgent operation 131 dams were designed and built to block and filter minor rivers and thereby diminish the movement of radionuclides. Three traps were built in the beds of the river Pripyat and of the Kiev reservoir to capture radioactive deposits. A drainage screen was built at the Chernobyl cooling pond. At present, plans have been prepared by the Ukrainian Ministry of Water Management to intercept the filtered waters from the cooling pond and also surface runoff water by means of an additional drainage system linking the open aquifers, and to use pumping stations to pump the water into the cooling pond, or to purify it by passing it through absorbent materials and then pouring it back into the river Pripyat.

There is also a plan to deal with the worst contaminated part of the flood-lands of the Krasnensk lake. Decontamination of the floodlands and of the standing waters in those areas will also involve processing and disposal of radioactive material. This is a radical way of preventing the seepage of radionu-clides into the river Pripyat and of improving the radiation situation along the whole Dniepr. In order to carry out this decontamination project the "Pripyat" scientific-industrial organization plans to enlist the aid of the "Ard" corporation of the USA, which has the necessary equipment and know-how (the "Scavenger" project).

It would be over-optimistic to hope for a substantial improvement in the near future of the radiation levels in the water ecosystems of the Dniepr basin in general, and of the Kiev reservoir in particular. Into this latter body of water flow all the streams and rivers which carry radioactive nuclides from the most heavily contaminated territories of the 30-km Zone and from the floodlands of the Pripyat. Eventually, all these waters go into the Black Sea which is directly connected with the Mediterranean.

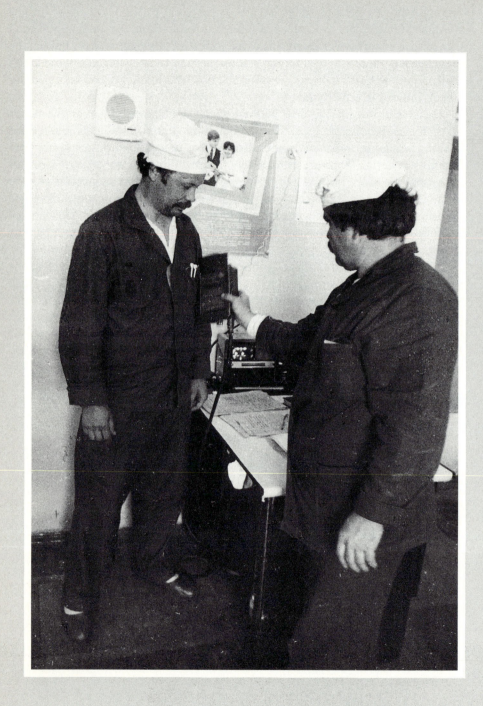

14. Risk, or How Safe is our Safety?

> "It is fruitless in years of chaos
> to seek a blessed end.
> Some are doomed to punish others
> and to suffer remorse –
> for the others – there is Golgotha."
> *B.Pasternak*
> *From the poem*
> *"Lieutenant Schmidt"*

In view of all the events since 1986, it is interesting to follow the evolution of radiation safety standards in the USSR. Analyses of all pertinent documents and directives that have been published in the last 30 years, and especially those of the post-Chernobyl era, show that there are still no scientifically grounded measures to protect the population from radiation exposure.

This fact has prompted endless public discussions and disharmony in the affected areas, and accusations by both scientific community and the press.

Taking into account the increased number of reactor accidents, particularly in this post-Chernobyl period, the absence of well-founded safety standards can lead to, and is leading to, irreparable damage to the health of present and future generations.

Preceding page, for workers in nuclear industry, possible radiation exposure and routine dosimetry are inherent. (Photo taken September 1986 in Chernobyl.)

14.1 A History of Radiation Safety Standards

Each sector of the economy, each "science" is governed by its own rules and regulations. From the thoroughness of regulatory documents it should be possible to judge the level of sophistication reached by any science, and to get an idea of the directions in which it is developing.

The law requires that those who draft rules for occupational safety not only ensure that those rules have a sound scientific basis, but that they be also subject to the approval of a number of ministries (the Soviet Ministry of Health, the State Construction Ministry, the Ministry of Internal Affairs), as well as approved by the Trade Union organizations.

Subordinate to the All-Union Soviet of Trade Unions (AUSTU) there are five scientific research institutes (in Moscow, Leningrad, Tbilisi, Sverdlovsk, and Ivanovo) that are concerned with occupational health and safety. On the basis of reports from these institutes the AUSTU obtains the agreement of the above Ministries to any proposed regulations. The procedure for obtaining agreement to proposed regulations is spelled out in a series of legal documents. We shall refer to *"Directives Relating to Procedures for Obtaining Approval of Standards and Technical Conditions from Trade Union Organizations"* RD 5O-III-86 AUSTU, State Standards organization, Moscow, 1987.

Remembering the two prerequisites – a scientific basis and the approval of all concerned bodies – let us review the process whereby regulations concerning radiation safety have been developed in the Soviet Union. This is a matter which now concerns practically everyone on earth, but most of all the inhabitants of the Ukraine, Byelorussia, and the Russian Federation.

We shall refer to opinions expressed by those who drafted the radiation safety norms which have been made public.

14.1.1 Thirty Years Ago

In 1961, following the definition in 1960 of maximum permissible levels of ionizing radiation in the Soviet Union, N. G. Gusev, now Professor of Technical Sciences, published his book *"Concerning the Maximum Permissible Levels of Ionizing Radiation"*. This book, as its preface says, is "part of the basic literature" on this subject.

> *Professor A. V. Lebedinskii, who is a full member of the Soviet Academy of Medical Sciences and the author of the book's preface, describes the focus and development of research in this field:*

The final aim of protection measures against ionizing radiation is the prevention of actual physical damage and the minimization of damage to the genetic structure of the population as a whole. The main focus of interest is not so much on the immediate effects of ionizing radiation as it is on the possible

long-term results, i.e., the shortening of the life-span, the reduction of reproductive capacity, the occurrence of leucosis and other malignant tumors The
genetic consequences of radiation show themselves in the descendants of irradiated individuals and may not be evident for many generations

The interest and, at the same time, the difficulty of this problem [of the
setting of radiation norms] arise from the need to bring together in the right
way all that we know about the biological effects of radiation. This requires a
careful synthesis of our knowledge in widely differing fields of science which,
at first glance, may not seem to be connected – for example, radiobiology and
meteorology, radiation hygiene and physics, biochemistry, and statistics etc..

*N. G. Gusev formulated this more concretely. Referring to the research
work of Professor Lebedinskii and of Yu. I. Moskalev,[1] he writes:*
In the work of these writers there is striking evidence of the great sensitivity of
living organizms to ionizing radiation. The various somatic effects appear both
in human beings and other animals at dose levels which are so low that they
are comparable not only to the permitted maximum levels but even to natural
background radiation.

Gusev states that the most serious problem is the one of small doses. He
writes not only of their effect on health, but, also about the difficulty of de-
tecting and measuring low-level contamination. For example, the permissible concentration of strontium-90 in the atmosphere is set at 0.00000000002
milligrams/m^3.

This is why, he writes (on page 175):
The inhabitants of London are less concerned about the 250 tons of soot and
dust that fall on each square kilometre of their city each year than about the
annual 0.0002milligrams/km^2 (3 millicuries/km^2) of strontium-90 that fall out
as a result of nuclear explosions. Apart from the difficulty of measuring such
small amounts, there is also the difficulty of localizing and containing them.

*The maximum permissible levels fixed in 1960 were based largely on
the recommendations of the International Radiation Protection Commission (IRPC, Munich 1959):*
Any change in the environmental conditions in which human beings are living
may lead to potentially harmful consequences. It is therefore assumed that
constant, prolonged exposure to ionizing radiation coupled with the effect of
natural background radiation poses a threat.

However, the human race is not able to forego completely the use of ionizing radiation and, therefore, the practical problem is that of limiting doses to
a level which will not bring unacceptable consequences for the individual or
the population as a whole.

---

[1] A. V. Lebedinskii, Yu. I. Moskalev *The Successes of Modern Biology* 1960

The term used for such a dose is the Maximum Allowed Dose or MAD. As far as the individual is concerned, the MAD is the dose accumulated over a long period or received in a single irradiation, which, as far as is known at present, is associated with a negligible probability of serious physical or genetic damage."

Gusev develops this idea in his book and gives a number of examples. He notes that when a thermonuclear bomb explodes more than 100 different radionuclides are formed. It is usually considered that the most dangerous of these are the long-lived isotopes strontium-90 and cesium-137, which have half-lives of 28 and 30 years, respectively.

Quoting UN data, Gusev writes that by 1958 the total worldwide fallout of cesium-137 had reached 7 MegaCuries, which is the equivalent of 25.9×10^{16} Bq.

According to the official Soviet figures sent to the IAEA after the Chernobyl accident, 50 MegaCuries of radioactivity were ejected from Block 4, 13% of which (6.5 MegaCuries) was in the form of cesium-137. Although this figure is disputed by a number of experts (who regard it as too low; see Chapter 1), we accept it here so that a comparison can be made of the relative official figures. However, we wish to point out that the 7 megaCuries of cesium-137 which fell on the whole surface of the world up to 1958 created more concern than the 6.5 megaCuries which later fell on a relatively small area of three Soviet republics.

According to the figures supplied by the Secretariat of the UN Scientific Committee on the effects of atomic radiation (New York 1960) the average measured concentration of strontium-90 in strontium units[2] in the bones of the population of the United States was (years – adults – children):

1955–56	0.07	0.56
1957–58	0.19	1.38

In the USSR the average concentration in children's bones for the years 1954-55 was 2.40 s.u..

In the USA at that time the level of strontium-90 contamination in the soil resulting from atomic weapons testing was 19 mCi/km^2. The USSR Health Ministry's standard for strontium-90 contamination since the Chernobyl accident has been 3Ci/km^2 (1.11×10^{11} Bq).

The Academy of Sciences of the Byelorussian republic gives the following figures:

Studies of the magnitude and character of the contamination of Byelorussian territory by strontium-90 show that, for the most part it, lies in the evacuation

---

[2] 1 s. u. is equal to approximately 3 millirads/year. For adults 1 strontium unit corresponds to a strontium-90 concentration of 0.15×10^{-12} Ci per 1 g of bones or 0.001 microCuries in the whole skeleton.

zone, where concentrations in the surface level of the soil exceed the permitted maximum. There are certain spots where the strontium-90 contamination level is more than 3 Ci/km^2, especially in the south of the Narovlya region. The level is $1 - 3$ Ci/km^2 in the regions around the towns of Khoiniki, Bragin, and Narovlya, around the settlement of Buda-Koshelevo, and in other places. In the rest of the Byelorussian republic strontium-90 contamination is lower than the permissible level.

True, but the "norm" referred to is 3 Ci/km^2, *which is 200 – 250 times higher than the level of contamination which so troubled the world in the early 1960s!*

Cesium-137 contamination caused just as much alarm. Gusev emphasizes that, unlike strontium-90, cesium-137 is a powerful source of gamma radiation and its accumulation in the soil may lead to a steadily increasing gamma-background of great penetrative power. Gusev calculated the expected dose for the citizens of Leningrad province resulting from global fallout. He arrived at the figure of 16 millirem over 30 years.

Using Gusev's data and methods it is possible to work out the dose received by the group 1 organs (gonads, bone marrow) in an environment where contamination reaches 15 Ci/km^2 – i.e., the level up to which inhabitation is permitted by the present Soviet Health Ministry standards. We arrive at a figure of 14.1 rem over 30 years, and 32.9 rem over 70 years. But this is only external irradiation.

Even in the 1950s research was done on the effect of radiation on the fetus and the resulting cancers in children under 10 years of age. British scientists established that every 20 mR increase the probability of cancer in children by 1%. We can conclude that when our scientists made the 1960 norms public they knew very well the dangers of radiation. They knew the seriousness of the problem and what had to be done about it.

14.1.2 The 1980s

Thirty years have passed since then. The science of radiation safety has obviously not stood still. Academician L. A. Il'yin of the USSR Academy of Medical Sciences wrote about the present state of this field in his article *"Radiation Hygiene: Problems and Tasks in the Light of the Decisions of the XXVIIth Congress of the Soviet Communist Party"*, which was published in 1985, not long before the Chernobyl accident.

L. A. Il'yin explains:

In radiobiology we must focus on the supremely important and difficult question of the so-called 'small dose effects'. It is to such doses, i. e., to total doses of 100 rem or less, or to doses of up to 5 rem/year – that most people are subjected, both in industry and elsewhere.

The study of the relationship between small radiation doses and the chances of cancer and birth defects is accompanied by enormous, practically insurmountable difficulties. The laws of statistics and the existence of spontaneous cancers and of naturally occurring genetic abnormalities complicate epidemiological research, as well as experiments on animals. It is obvious that the removal of this difficulty requires research into the fundamental nature of cancer and the mechanisms of carcinogenesis.

The precise definition of the risk arising from small doses would make it possible to optimize the protection of the general population and of specialized staff, by applying it in exactly the right form and the right circumstances.

At first glance one might conclude that the problems have not changed. There has been no great progress. The circumstances, however, have unfortunately taken a sharp turn for the worse. Instead of the strontium-90 and cesium-137 contamination resulting from nuclear testing which could be measured in tens of milliCuries to the square kilometer, and which caused such an alarm in the 1960s, we are now dealing with strontium-90 levels of up to 3 Ci/km^2, while cesium-137 levels of up to 15 Ci/km^2 (even up to 40 Ci/km^2 according to Yu. A. Izrael) are regarded as safe. In other words, levels have risen by two or three orders of magnitude, yet the world at large is not worried.

How are these two facts to be reconciled: the stagnation of the science of radiation safety and the acceptance of this by scientists, even as the earth becomes increasingly contaminated by radionuclides? Clearly, the answer will not be found in the depths of radiation medicine, but in other sciences more easily affected by social changes – in politics, philosophy, and logic.

Science was faced with a question of priorities: whose needs came first – those of nuclear power or those of those of the population? Public health, or the health of the nuclear industry? Which way the choice went can be judged by anyone who analyzes the development of our radiation safety regulations.

14.2 Current Radiation Safety Standards

The basic legal framework for radiation protection in the USSR is provided by *"Radiation Safety Norms – 76/87"* and *"Basic Health Regulations 72/87"*.[3] RSN-76/87 was not the first such document. Previously there had been RSN-76 (1976), RSN-69 (1969), and MPL-1960 (1960).

> *The oldest of these, MPL-1960, was drawn up on the basis of the recommendations of the International Radiation Protection Commission (IRPC-1959), in the preamble to which we find:*

---

[3] For further details on radiation safety terminology the reader is referred to Appendix A.

The aim of protection from ionizing radiation is to prevent physical damage and to keep to the minimum any damage to the genetic structure of the population as a whole. The most serious physical consequences of irradiation include leukemia and malignant tumors, loss of reproductive capacity, cataracts, and shortening of the life-span. Genetic damage shows up in the descendants of irradiated people and cannot be discovered for many generations. Even if only isolated groups are irradiated the genetic effects can spread through the population as a result of marriages between irradiated and non-irradiated individuals.

Thus, the IRPC notes that, in the case of genetic damage, there is no dose threshold below which there will be no effect – that is, the genetic effect is directly proportional to the dose. However small the dose, it will have an effect.

A linear correlation between dose and genetic effect has been established in all experiments carried out up to now on viruses, microorganizms, multicellular plants, and animals.

> *Commenting on the IRSC Recommendations and on MPL-1960 in 1961, Gusev wrote (p. 105):*

In contrast with the situation in a number of capitalist countries, in the USSR such norms and rules are not mere guidelines, but have the force of law and are binding on design organizations, as well as the corresponding enterprises, institutes, ministries, and other institutions.

Thus, MPL-1960 was a document that was intended to protect, not only professionals, but also the general public from radiation. A comparison of the corresponding sections and rules in all the documents from MPL-1960 to RSN-76/87 makes this clear.

All four statements define three categories of persons: A, B, C (Appendix A), each with a differing proximity to radioactive substances. All four statements establish allowed annual dose levels for several groups of tissues with varying radiation sensitivity. RSN-76 and RSN-76/87 contain the same limiting values. However – and this is most telling – only MPL-1960 sets limits on the exposure of Category C individuals: the population at large. In RSN-69, Section VII *"Irradiation of the population as a whole"* it is stated:

§ 1. The genetically significant dose of external and internal irradiation (including the dose from global radioactive fallout) received by the population as a whole from all sources must not exceed 5 rem over 30 years. This total dose does not include the dose given in medical treatments or natural background radiation.

The allowed genetically significant dose of 2 rem/30 years for Category C is 1.5 rem of internal irradiation and 0.5 rem for external irradiation.

§ 6. In the event of an accident, the Central Epidemiological Department of the USSR Ministry of Health determines, based on the scale of the accident, temporary permissible levels for irradiation and the ingress of radioactive substances into the body. When this is done the genetically significant dose for the population of the country as a whole must not be exceeded. It may be exceeded only in certain individual cases among the population of a given republic.

In RSN-76 dose limits for Category C are not fixed. Instead, Paragraph 6 "Irradiation of the population (Category C)" is introduced.

§ 6.1 Limits on the irradiation of the population (Category C) are fixed according to the chances of long-term effects and genetic consequences. The regulation and verification of the irradiation of the Category C population is the responsibility of the USSR Ministry of Health.

§ 6.2 In the event of a radiation accident the Central Epidemiological Department of the USSR Ministry of Health determines, based on the scale of the accident, temporary permissible levels of irradiation and temporary limits on the ingress of radionuclides into the body.

In RSN 76/87, dose levels for the population are not fixed. Section 6 "Irradiation of the Population", and Section 7 "Irradiation of the Population Due to Accidents" are introduced:

§ 6.1 Limitation of the irradiation of the population (Category C) is achieved by monitoring of the following: the radioactivity in the environment (water, air, food, etc.); technological processes where contamination with radionuclides is possible; doses from medical irradiation; background levels raised by the use of certain building materials, by chemical fertilizers and by the burning of organic fuels, etc.. Limitation of the irradiation of the population (Category C) is also partly defined by the dose limits for Categories A and B which are established by this document.

The nature of regulation or monitoring is defined by BHR-72/87 and other regulatory acts approved or agreed on by the USSR Ministry of Health as required by law.

§ 7.1 In the event of an accident in which people outside the protective zone (of a nuclear power station) are affected by radiation, the dose limit may be raised. All practical measures must be taken to keep irradiation and environmental contamination to a minimum in accordance with the requirements of BHR-72/87. Any territories where the indicated levels can be exceeded are part of the radiation accident zone.

§ 7.2 In taking the appropriate measures, given the scale and character of the accident, the USSR Ministry of Health may set basic dose limits for the population and permissible levels while drawing up protective rules which will permit normal life to continue in the contaminated areas.

Thus, as far back as 1976 the Soviet population was rendered defenseless in the face of nuclear power station accidents. All norms and limits were abolished, or rather forgotten. The Ministry of Health acquired the right to set the norms for Category C. However, mention of "possible long-term effects and genetic consequences" is retained in Paragraph 6.1. In RSN-76/87, therefore, this relic of the old regulatory mechanism was deleted.

An amazing amalgam of hypocrisy and bureaucratic jargon! What have dose levels for Categories A and B to do with the rest of the population, with women and children? What can be set for the population in the event of an accident? Referring to BHR-72/87, one is struck by the fact that there is not a word there about the regulation and monitoring of the irradiation of the population following an accident – and, of course, no sign of any numbers.

One cannot help thinking, when reading RSN-76/87, that science must have proved that radiation has an effect on Categories A and B, but that Category C (the population) is invulnerable. Apparently, this is the scientific basis for the present legislation on the protection of the population from radiation. There is no other explanation for the document issued in 1988 by the Ministry of Health under the title *"The lifetime dose limit established for the population of the monitored areas of the Russian Federation, Byelorussia, and the Ukraine contaminated as a result of the accident at the Chernobyl nuclear power station."* In Paragraph 1 of this statement the lifetime dose limit for external and internal irradiation is set at 35 rem. But 35 rem over 70 years is 0.5 rem per year, i.e., the norm for Category B (see MPL-1960, RSN-69, RSN-76, and RSN-76/87)! Then comes a paragraph which simply amazes by its cynicism:

§ 3. It is appropriate that the population of the monitored (contaminated) areas should be grouped with persons of Category B given the basic definition of the latter category, i. e., that they are persons who are exposed to the effects of radiation and are consequently under medical observation.

But even the "35 rem over 70 years" limit was not the final word on regulations. In September 1988 the Ministry of Health issued another document called *"The dose limit fixed for 1988 and 1989 following the accident at the Chernobyl Nuclear Power Station."*

§ 1. The dose limit for total external and internal radiation is set to 2.5 rem/year.

So now we have a situation where four regulatory documents are in force: RSN-76/87 and BHR-72/87 with no norms set for the population (Category C), the Ministry of Health document of November 22, 1988 setting a limit of 35 rem over 70 years for Category B for 1989, and equating the general population with this category, and finally, the Ministry of Health document of September 1988 with a limit of 2.5 rem/year for 1988 and 1989.

Before arriving at any conclusions, let us consider the question of the harmonization of regulatory documents prepared by the USSR Ministry of Health. We have no information as to who drew up and approved MPL-1960. RSN-69 was drawn up by the National Radiation Protection Commission (NRPC) and approved by the Chief Medical Officer of the Ministry of Health of the USSR

P. N. Burgasov. RSN-76 was prepared by the working group of the NRPC and submitted for scrutiny and approval to the NRPC, itself part of the Ministry of Health. It, too, was approved by Chief Medical Officer of Health P. N. Burgasov. RSN-76/87 was prepared by the working group of the NRPC, submitted for scrutiny and approval to the NRPC itself in the Ministry of Health and finally approved by Chief Medical Officer of Health of the USSR G. N. Khlyabich.

No one outside the Ministry of Health had any hand in the preparation and acceptance of this document, which establishes norms for the protection of the whole population from radiation.

It is simply amazing that the All-Union Soviet of Trade Unions, with all the institutes and technical inspectors that it has at its disposal, with its reputation as the watchful guardian of the workers' interests, "forgot" about the norms relating to radiation safety. It is true that AUSTU gave its approval to BHR-72/87, but these, after all, are only "Basic Health Regulations" and not "Radiation Safety Norms". Moreover, no one noticed that BHR-72/87 was drawn up in such a way as to accept the provisions of RSN-76/87, which had never been submitted to the AUSTU. The BHR also deal with Categories A and B, but there is no mention at all of Category C.

One is justified, therefore, in concluding that the USSR has absolutely no scientifically based laws ensuring the protection of the population from radiation.

The neglect of its duties by the AUSTU has led to a situation in which the standards for women and children are several times higher than the levels to which nuclear power workers are subjected: in 1986 – 10.0 rem; in 1987 – 3.0 rem; in 1988 – 2.5 rem; in 1989 – 2.5 rem. Meanwhile, the actual average dose of a nuclear power worker is about 1 rem/year.

Obviously, there is only one way forward: the Ministry of Health must be compelled to draw up *"Radiation Safety Norms for the Population (Category C) of the USSR"*. Then it must obtain approval for the document according to the established procedure and have it passed into law. These standards must be as detailed and specific as those for Categories A and B.

In the conditions now existing after the Chernobyl accident, and with the situation in the atomic energy industry being what it is, such regulations are essential if we are to avoid preparing the way for an irreversible deterioration of the health of this generation of Soviet citizens and of future generations.

To fill the gap while the necessary regulations are being passed, it is vital that temporary norms be issued. In particular, there must be a ban on the needless irradiation of an innocent and unsuspecting population.

Fig. 14.1. Risk assessment with regard to radiation is impossible without conscientious measuring and monitoring (Photo taken in May, 1986.)

г. Чорнобиль.

15. The Legacy of Chernobyl

"An analysis of the present position regarding the Chernobyl nuclear power station shows that we have somehow managed to reassure ourselves. We have managed to feel confident about the way things are going — everything seems to have been done correctly. Programs are drawn up, decisions are made. From time to time something gets done. The most frightening aspect of all this is that an awful lot of talking goes on and the responsibility for various matters is spread around among ministries and organizations. You can already feel that everyone is coming to terms with the idea that the Zone is with us for good. But the radioactive contamination is still spreading and the radioactive ruins of Block 4 have not been made safe as required by the RSN-76 regulations. People are becoming concerned about the slow progress towards the solution of all the problems which resulted from the accident. We are, in fact, still in the same situation as we were at the time of the accident, except that it is moving more slowly and in a different ecological direction.

"That is why we must get things organized again as we did in 1986.

"Chernobyl must take priority over all other questions — whether in the Ukraine, in Byelorussia, or in Russia ..."

Academician Viktor Baryakhtar,
Vice President of the
Ukrainian Academy of Sciences,
Chairman of the
Ukrainian Academy of Sciences
Chernobyl Emergency Commission

Preceding page, a flag has been hoisted on top of the ventilation chimney of Block 3, symbolizing "completion" of the rectification work. Yet, centuries will pass before true rectification can be considered complete. (Photo taken October 1, 1986.)

Five years have passed, but life in the Zone has not changed. Ignoring all good sense, three blocks of the Chernobyl power station continued to operate until 1991, when the Ukrainian Supreme Soviet decided to close them down. The most dangerous of these three blocks, Block 1, has already been powered down.

The "Kombinat" project established in late 1986 has been transformed into the production union "Pripyat". Tens of thousands of people still work in the Zone, in 15-day shifts every month. The importance of the output of this organization is not so obvious, but the radiation conditions do certainly affect the health of the employees. The enterprise costs the national budget up to 500 million roubles per year.

We have seen that the Chernobyl accident is unique in several aspects in terms of geographic scale and degree of contamination. The Zone should be declared an international research center for the investigation of the myriad of problems left over from the catastrophe, including those of decontamination, reprocessing the millions of cubic meters of radioactive discharges, and studying the influence of the radionuclides on the surrounding ecosystem.

Without a doubt, the most important problem is the well-being and monitoring of the health of the inhabitants of the huge polluted regions: the investigation of the influence of small doses of 0.3 to 0.5 rem/year, the dosimetry of radionuclides, and the development of methods of treatment.

In this chapter, attention is focused on the current status of the government Chernobyl rectification program. At first glance, it may appear to be primarily of interest to the Soviet reader; however, the problems touched upon concern every country with atomic power stations in use, or neighbors of such countries. Reactors are "burned out" after around 30 years. Then the remains of the power station have to be isolated or otherwise neutralized. The latter involves technological problems of the same type (but, thankfully, on a smaller scale) than the ones faced at Chernobyl.

15.1 Comments on the Chernobyl Rectification Program

1) The Program presents no comprehensive assessment of the situation in the contaminated territories. There are no forecasts of social consequences. There is no estimate of losses caused by the accident (unofficial estimates suggest losses amounting to 200 billion roubles by the year 2000). The Program has no legal status. There is no clear statement of aims or sequence of operations, of of the requisitioning of the necessary scientific, human, financial, and material resources.

To date, the government has actually accomplished only the following:
– the containment of the most powerful sources of radiation in the accident zone;
– the restart of three of the Chernobyl reactors;
– the resettlement of the Chernobyl staff to the town of Slavutich.

2) The main question is not properly settled, i.e., what becomes of the people? The population, the "rectifiers", the soldiers? The Program's purpose seems to be only to:
– soothe public anxiety;
– exonerate the real culprits;
– minimize expenses.

3) The Ministry of Health's "35 rem over 70 years" guideline can be viewed as a decision about how many people should be allowed to die.

4) It is wrong to give, instead of clean food, only the money to buy clean food. Clean foodstuffs are not brought into the contaminated areas, at least not in sufficient quantities.

5) There should be no talk about privileges for rectifiers and local people. Privileges were invented by the bureaucracy as a mechanism by which it could lift itself above the rest of society. The question is one of legal compensation for damage done to health, life, livelihood – not of handouts. This just compensation must take various forms: medical treatment, financial help, jobs, pensions, assistance in the home.

6) There are no criteria as to what constitutes decontamination, the scope of the decontamination task has not been established, and already the army is spending 1 million roubles a day on this work.

There are no plans to decontaminate the villages, lakes, rivers or the Kiev reservoir (although radioactivity is already building up in the rice fields around the mouth of the Dniepr).

Only the Byelorussian Republic is planning to construct disposal pits and even there the volume of the planned pits totals only 300 000 cubic meters. This is sufficient for the decontamination of only 30 000 hectares if a 10 cm layer of soil is removed. However, there are 3 million hectares on which the contamination level is 15 Ci/km^2. This does not include the 30-km Zone and the central patch. Decontamination of one square kilometer yields 140 000 tons of waste. Readiness to carry out intelligent, large-scale decontamination operations is totally lacking in all quarters.

There is no ban on the burning of contaminated peat. The radioactivity accumulated in the peat is released into the environment along with the smoke and ash.

7) There is no attempt to further search for and collect "hot" particles which are widely scattered.

8) No proper radiation maps have been made available to the workers in the contaminated territories, and there are no plans to remedy this. The maps supplied by the State Meteorological Service are very large-scale. Work with such maps is necessarily very imprecise – five to eight times less precise than with the 2-kilometer (1:200 000 cm scale) maps which are really needed. The maps give no information about the thousands of (temporary) pits where radioactive debris has been buried (neither has there been any information about what the authorities plan to do with them). In the 30-km Zone alone there are 800 such temporary pits.

And lastly, there are no hydrological maps or forecasts of contamination spread.

9) All the agronomic measures are based on the "35 rem over 70 years" guideline. This permits the production of food on land where contamination levels reach 80 Ci/km^2 and more. Before the accident only products which were intended for further processing were permitted to be grown on land where the radioactivity level was between 3 and 4 Ci/km^2. There was a complete ban on the cultivation of land where the level reached 4 Ci/km^2.

1520 Byelorussian settlements produce milk which contains more than the temporarily permitted level of contamination, i.e., 1×10^{-8} Ci/liter. Bones which contain 8 to 90 times the permitted amount of contamination are ground up as food additives. Only this year (1990) has the Byelorussian Republic decided to stop food production on land where contamination is at the 40 Ci/km^2 level.

15.2 Suggestions for Modifications and Alternatives

The Program should be revised and governed by two objectives:

1) First by conducting *a complete survey* of the effects of the Chernobyl accident and of its possible local and global consequences, and by clarifying all consequences such as:

   - technological
   - ecological
   - medical
   - sociopolitical
   - economical
   - psychological.

These should be divided into two groups of predictable and immediate consequences, and unpredictable, but probable, long-term consequences. (It is impossible, of course, to fully assess the unpredictable consequences – but it is necessary to expect them.)

It is necessary to assess the harm done if no actions were undertaken, and this would help to put the costs into proper perspective.

In order to establish the scale of the accident it is necessary to have available accurate data on the emission and distribution of contamination, on radiation doses and their consequences. Without the collection and analysis of such data the program has no rational basis. Some data has been collected over the last 3 years, including data about Block 4, and it shows how much radioactivity we are dealing with at present and how it is distributed.

A large area around the upper reaches of the Dniepr (100 km^2) is contaminated with a huge amount of radioactive material (500 000 Ci). In this area are the 800 debris disposal pits, disposal areas for low-level power station waste, the Sarcophagus, the Spent Fuel Store (containing not less than 20 billion Ci), and the three RBMK reactors still working at Chernobyl. An accident involving any one of these could turn the Dniepr into a river of death and force the evacuation of 10 million people now living in the Ukraine. We must realize the enormous threat which radiation poses to every living organism on the earth. (To poison all the water drunk in a year by the population of the USSR takes only 100 g of cesium-137, or 10 000 Ci.)

The results of the work must be published.

2) The creation and implementation of *a comprehensive rectification program* with all the necessary scientific, organizational material and technical backup and long-term financial provision at international, national, and local levels.

In order to create a Comprehensive Program it is necessary to involve, not only the Council of Ministers and the Academy of Sciences of the Ukraine, but also informal public organizations. Emphasis must be placed on the measures which can be definitely carried out given the existing material, technical and economic resources.

The Program must define the task for each link in the chain, from the scientific study of the problems right through to implementation and to the assessment of the results achieved.

Specific persons should be responsible for the implementation of each component of the Comprehensive Program. The monitoring of the progress of the rectification operations by the public is essential.

Technological Problems

1) The radioactive material ejected from the reactor over wide areas must be neutralized. Only time can solve this problem completely. There is no technological process which can destroy radioactivity, the only hope is to collect the contaminated material and put it into safe, leak-proof, long-term storage. The special plant (the "Vector" factory) in the 30-km Zone will not solve all the problems. No country in the world has experience in processing, on such a scale, millions of cubic meters of contaminated material. Unless we find some way of turning temporary dumps into permanent storage units, which does not involve transporting their contents over long distances and which makes use of simple techniques, we can be sure that we shall have to spend huge sums on the development of new processing technologies. The problem will drag on for decades.

2) What are we to do with the spent fuel from the Chernobyl station (30 000 fuel rod assemblies containing tens of billions of Curies of radioactivity) – the USSR possesses neither a suitable processing plant nor suitable transportation facilities? The maximum life of the Spent Fuel Store has not been precisely determined. Judging by previous experience of the storage of spent fuel in storage ponds at nuclear power stations, it is possible to forecast that leaks will appear within 10–15 years from the beginning of the storage period (1986).

It is also difficult to imagine how the highly radioactive waste from the "Vector" plant is going to be stored, given that in the USSR highly-radioactive waste is kept in containers which range in capacity from a few tens to a few hundred cubic meters. Because of the potential danger represented by this method of storage (as shown by the "discharge" at Chelyabinsk in 1957), it must be considered as only a stopgap measure.

Let us assume that the period of storage in such containers should be no longer than 20–30 years. The problems involved in the consolidation of such waste have already been studied for 30 years without any practical outcome.

Unless this problem is solved it will not be possible to dismantle the Spent Fuel Store, nor the Sarcophagus, nor any of the Chernobyl reactor blocks.

Even if the necessary plant were to be created, its presence on the upper Dniepr would represent a grave potential threat to the region.

3) Alternative energy sources must be developed. This is certainly an international challenge.

Ecological Problems

We have yet to move beyond talking to a stage of making recommendations. It is essential to develop a scientific and practical base in the Ukraine from which to tackle our ecological problems.

1) The first task is to protect the Dniepr basin from further radioactive contamination. Then a practical decontamination program for the Dniepr must be worked out and implemented.

2) The Zone should be restored to its pre-accident state. Even in theory this is an impossible task. Some kind of reasonable compromise between ecology and economy will clearly have to be sought. However, an assessment of the possibilities must be made immediately and the public must be aware of them.

Medical Problems

1) The population must be protected from the radiation released by the Chernobyl accident. The protection of the population is impossible in the absence of legislation regulating ionizing radiation sources applicable to all industries, the medical services, and transportation. Protection of the population from radiation must be required by law and secured by a proper system of inspection. Meters must be made available to enable individuals to monitor their own doses.

Appropriate articles must be inserted into the criminal code making it an offense to allow excessive irradiation of people.

2) It is necessary to abandon the faulty practice of the USSR Ministry of Health in determining the average dose of radiation received by the citizens of Kiev or Byelorussia. The dose is not evenly distributed and there is no such thing as a safe dose. It is important to know each person's individual dose.

3) It is not sufficient to compile only one main national register of people who have received doses of radiation as a result of the accident. The experience of the last few years has shown that such a register is of limited usefulness because it is unwieldy. The lack of ongoing communication with the regions where these people live leads to continual deterioration of the register. As much as 10% of the stored information becomes outdated each year as people move, change their names or die.

The Ministry of Health must comply with the requirement laid down in Paragraph 7.3 of the RSN-76/87 (Radiation Safety Norms) regarding the assessment of individual doses of external and internal radiation, and for maintaining the registers at republic, provincial, and regional levels.

(In Japan the radiation victims of the atomic bombs held special identification cards and a sophisticated monitoring system kept them under the constant scrutiny of state medical and social services.)

4) There must be an effort to develop the scientific base of radiation medicine in the Ukrainian republic. A specialized resource center for medicine and diagnostic equipment should be set up. A system of specialized polyclinics, hospitals, and treatment centers should be established in the Ukraine under the direct control of the Ukrainian Ministry of Health and the Ukrainian Association of Trade Unions.

There should be permanent, active international collaboration with medical institutions which have developed sophisticated radiation medicine.

5) A National Radiation Protection Commission (NRPC) must be created within the Ukrainian Health Ministry. The NRPC should have the power to enforce standards and regulations concerning radiation and it should assume a supervisory role.

6) All medical institutions in the Ukrainian republic which are at present administered by the Central Ministry of Health of the USSR should be transferred to the Ukrainian Ministry of Health (together with their budget allocations). This includes the All-Union Radiation Medical Research Center in Kiev (AURMRC).

Sociopolitical Problems

1) The existing censorship which prevents the dissemination of the truth about the Chernobyl disaster and its consequences, and about nuclear energy in general must be abolished immediately.

2) The bans on the dissemination of ecological information must be lifted. Ignorance of the dangers and failure to act to prevent them may lead to grave consequences, even to the degeneration of the population.

3) It is necessary to foster the growth of independent public expertise in ecological matters embodied in the organizations "Zeleny Svit" and "Narodny Rukh",[1] and to give the public a right of veto in related matters.

---

[1] "Zeleny Svit" or "Green Light" is the informal organization of the Greens in the Ukraine, it deals with ecological questions. "Narodny Rukh" is a political organization supporting the perestroika (founded towards the end of 1989).

4) All data relating to the disposal of contaminated material in the territory of the Ukrainian republic must be published.

5) All proposals and decisions of central Union agencies in the field of nuclear power must be approved by the Supreme Soviet of the Ukraine and by the provincial and local Soviets.

6) The 30-km Zone and all installations within it must be brought under the jurisdiction of the Council of Ministers of the Ukraine, with overall supervision to be exercised by the Ukrainian Supreme Soviet. The Ukrainian Council of Ministers should have the sole right to decide on matters relating to use of land, to use of woodland, and to construction in the 30-km Zone.

7) The Ukrainian Academy of Sciences should have the exclusive right to plan scientific research projects and to permit them to be carried out in the 30-km Zone by Soviet and foreign scientific organizations.

8) All organizations and enterprises, in fact the whole population, must become involved in the continuing effort to bring the contaminated areas back to life.

9) There is a need to create a public coordinating committee, based on the examples of the Zeleny Svit and Narodny Rukh organizations, to supervise and organize public efforts. This committee should be made responsible for a new International Fund which would provide the resources needed to accomplish the tasks that have been set. The Fund would receive voluntary contributions from Soviet and foreign citizens ("The Regeneration Fund").

10) The republic's program "Homes 2000" should be considerably expanded to provide a far greater number of dwellings to receive evacuees from contaminated areas.

Economic Problems

1) The Chernobyl reactors must be shut down permanently. The RBMK design used in the first and second phases at Chernobyl (Blocks 1 and 2) fails to meet the requirements of the SRNPS-04-74 (Safety Regulations for Nuclear Power Stations) and of GSG-73 (General Safety Guidelines), which were in force while it was being developed. The large number of serious infringements of the regulations has been discussed explicitly in Chapter 4. The accident in Block 4 should have been expected, given the serious design deficiencies of the reactor, including those in the control and protection system, as well as in the poor quality of the operational documentation.

It is no accident that no full, mutually agreed-upon report on the accident has yet been issued by the Chief Science Officer (the Kurchatov Atomic Energy Institute), the Chief Designer (Scientific Research and Development Institute of Energy Technology), the General Planning Supervisor (the Atomic Energy

Planning Institute), and the Byelorussian Atomic Energy Research Institute. The State Atomic Energy Inspectorate insisted back in 1989 that such a document must be produced.

Because the RBMK reactors lack adequate accident localization systems and represent a great potential danger, it is essential that a timetable for their withdrawal from service should be immediately included in the Rectification Program. Their power output can be replaced by that of small local fossil fuel stations, which are now being built, or by other power sources.

2) The real long-term cost of the rectification operations (up to the year 2000 and beyond) should be determined. Expenditures on rectification work should be a separate item in the budget of the Ukrainian republic, and contributions from Central Union funds should be allocated.

3) The Supreme Soviet of the Ukraine must insist on a radical change of attitude about the Chernobyl problem on the part of the government of the republic. It is essential that the Ukraine should have its own approach, quite different from the activity of the USSR Ministry of Health, the State Meteorological Service, and other agencies which take the attitude that the damage done by the accident has been rectified and that there is no need to expect further serious consequences. It is clearly impossible to change the opinion of these people who "advise" the Union government – but it is essential not to be associated with them. The Ukraine has enough scientific and engineering resources to work out intelligent and responsible ways of solving the problems arising from the Chernobyl accident.

Psychological Problems

It is vitally necessary to use all possible means to make people understand the seriousness of the situation and to realise how tragic the consequences of the Chernobyl accident may be unless all possible efforts are made to rectify them as soon as possible.

15.3 Outlook

In this book, we have examined the results of a single accident. Its horrendous dimensions must force us to weigh the pros and cons of our energy-producing technologies.

In theory, nuclear technology offers the possibility of producing cheap and ecologically clean energy (provided all problems with the radioactive waste are solved). On the other hand, fossil-fuel and hydroelectric plants are not exactly environmentally neutral, especially regarding the "greenhouse" effect, damage to the ozone layer, or the massive territories flooded by dammed rivers or those transformed into deserts downstream.

In addition, attention has to be drawn to the fact that – due to radioactive nuclei found in natural coal – coal-burning plants release per annum amounts of radiation that are estimated by some to be even higher than those minimal quantities that may be given off by nuclear power plants.

The acute dangers of these conventional technologies, meaning the probability of immediate injury or death in the case of accident, might be as high as that for nuclear power stations. But the spatial and temporal dimensions of the effects of a comparable accident at a nuclear plant are several orders of magnitude larger. Potentially, a single accident at a nuclear power station may cause so much damage that the theoretical advantages of this technology are grossly reduced.

Thus, we are confronted with questions of methods and principle. With regard to methods, we must ask whether it is possible to guarantee, with reasonable effort, a given safety level that satisfies the demands of society and to achieve this at acceptable and bearable costs. This is equally true for still alternative means of energy production such as wind, geothermal, and solar power. As regards principle, crudely speaking, this is a decision between "cheaper energy at a larger risk" and "expensive energy with increased safety".

The most important conclusions are these: The area of the territory that would be affected by all aspects of the chosen energy technology must not exceed that of the jurisdiction of the decision-making bodies. Decisions involving potentially dangerous technologies must not be made without an honest, fair, and qualified public discussion.

16. Chernobyl

Reflections and Photographs

Photographs taken

by **Aleksandr I. Salmychin** [1]

---

[1] A.I. Salmychin took part in the rectification work within the 30-km Zone since August, 1986. At present he is working in Pripyat as a member of the emergency team ("Spetsatom") created to deal with situations similar to the one in Chernobyl – should they arise again.

The town of Pripyat is spread along the banks of the river Pripyat in the northern part of the Ukraine. This aerial view, *top,* shows the functional design of the highrise apartment buildings and a glimpse of the picturesque river. A bit further south, the river Pripyat – together with the Dniepr, Uzh and other rivers – feeds into the Kiev reservoir. The first city at the southern tip of the reservoir is Kiev, the capital of the Ukraine, located about 130 km from Pripyat. The waters eventually flow through other reservoirs to the Black Sea, supplying a multitude of towns and cities along the way.

Quite a few of the 40 000 inhabitants of Pripyat were employed at the nearby Chernobyl Nuclear Power Station, making sure that the "tamed atom" stayed tame. The power plant is situated close to the small town of Chernobyl at the northern end of the Kiev reservoir.

One night in April, 1986, a nightmare came true: A sudden explosion at Block 4 shattered the walls, lifted the roof, tilted it, and crashed it down into the reactor hall. This would have been a dreadful event even on a conventional scale – with fire, steam explosions, hot water, and falling steel and cement. But in the case of a nuclear reactor this constitutes a global disaster. Radiation kills people in the immediate vicinity; unchecked and leaked into the atmosphere or into aquifers, it can poison people and environment over huge territories.

Because of high radiation levels, airplanes and helicopters were the best means to obtain an overall picture of the shattered reactor.

In later stages helicopters were used not only for reconnaissance, but also to hinder the further spread of radiation by spraying polymer liquids onto the open reactor where they formed a thin film that restricted the free movement of dust. This dust, carried by the wind, was an additional source of radiation pollution besides the direct emissions of "hot" particles from the reactor cavern, which spread radiation over large parts of the Soviet Union and Europe.

The dangerous radiation levels forced the total evacuation of the town of Pripyat. In this photographically foreshortened view of Pripyat (*below right*), the photographer illustrates that the nuclear power station looms menacing over a ghost town.

The children of Chernobyl were shipped off to Young Pioneer Camps.

Decontamination in Chernobyl. Washing the radiation down the drain.

Since we cannot sense radiation, it needs to be measured. This does not, of course, eliminate the danger, but helps to minimize it by dictating the level of necessary precautions. Contaminated soil and debris are collected and buried, then the earth moving equipment, all vehicles, structures, and people must be decontaminated.

The trucks and tanks called in for the heavy construction and decontamination work have to be washed repeatedly. Nevertheless, many of them become so contaminated that they have to be buried in hurriedly dug "tombs" for nuclear waste.

There is also the serious danger of contaminated cooling and wash water seeping into the groundwater, thus poisoning all waters further downstream. Hence, dams and other barriers are built to confine the polluted material to the initial danger area.

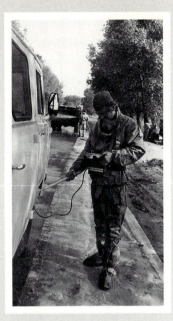

To this day, attempts to decontaminate soil and water systems continue.

In the above right photo a soldier futilely, in fact – ridiculously, attempts to hose down the huge Kiev reservoir with decontamination fluid from a tanker truck.

Directly after the accident some of the soldiers who were called in to help were housed in flimsy tents pitched amid high radiation levels.

Precautions like special protective clothing (*bottom left*) were of utmost importance during the "liquidation of the consequences of the accident". However, such special gear can provide only limited protection, particularly, when radiation is extremely high.

Not only hundreds of thousands of workers and soldiers came to Chernobyl after the accident, but also various commissions paid short visits to the site. The *bottom right* photograph shows the Chief Designer of the RBMK reactors used at Chernobyl, Academician A. P. Aleksandrov (from left to right: Ignatenko, Umanets, and Aleksandrov).[1]

---

[1] Ignatenko was the director of the factory "Kombinat", created in September, 1986, in Chernobyl to deal with decontamination work within the 30 km Zone. Umanets was appointed as director of the Chernobyl Nuclear Power Station in February, 1987.

The sign reads:

"Radioactivity
DANGER ZONE!
Cattle grazing, cutting of hay,
picking of mushrooms and berries
PROHIBITED!"

These officials are members of the Government Commission
on the "Liquidation of the Consequences of the Accident" (left to
right: Voznek, Chairman Shcherbina, and Aleksov).

Aerial inspection from helicopters was extremely important for evaluating the situation and deciding on further measures. General N. Tarakanov, responsible for the cleanup operations on the roof of Block 3 contemplates the disaster from the air. How many victims will this battle still claim?

The Directors of the Task Force for the Rectification of the Consequences of the Disaster ("Likvidatsiya posledstvii avarii" – LPA) within the Special (10-km) Zone hold an on-board discussion of their observations during this inspection of the danger zone (left to right: Engineering Director, V. Golubev and the Scientific Director, (author) V. Chernousenko).

After a flight to inspect Block 4: the Directors of the Cleanup with the crew of their helicopter.

This very crew became a victim of the rectification work the following day, October 10, 1986. They were flying very close over Block 4 when the tailrotor of their helicopter struck a chain dangling from the "Demag" crane being used to build the "Sarcophagus". The helicopter crashed about 5 meters from the Sarcophagus. The crew was killed.

The ventilation chimney of Block 3, one of the most dangerous high-radiation areas. The scattered bits of material are highly radioactive graphite from the reactor core which the explosion deposited on the chimney platforms and onto the roof of Block 3.

The extremely high radiation fields on the roof of Block 3 fouled the electronics of the robots and brought them to a standstill. Since politicians insisted the work could not wait, it was completed by hand. To keep the radiation doses within limits, work shifts of 15 to 90 seconds were scheduled for each individual.

Only rather rudimentary protective gear was available.

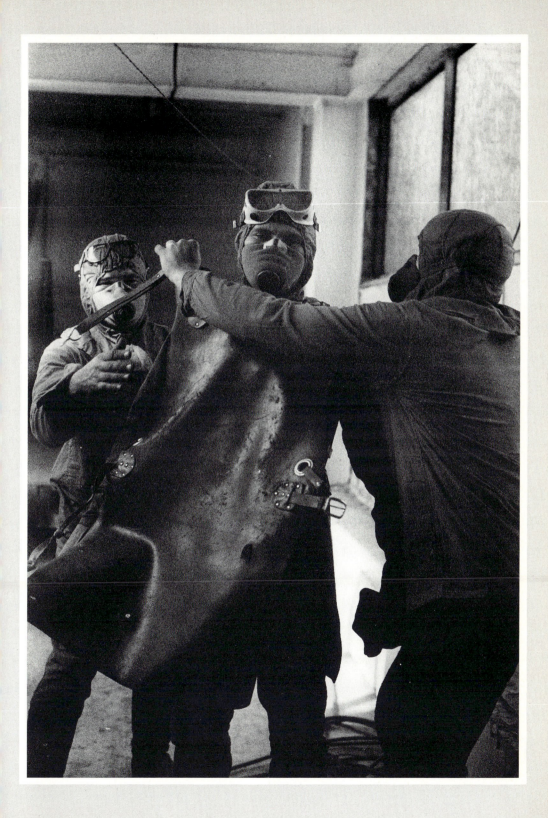

Both the chimney platform and the roof of the neighboring Block 3 were cleared of radioactive debris which was thrown into the cavern of Block 4. Simultaneously, a containment structure was being built around the remains of Block 4. This is a view of the completed "Sarcophagus".

To complete the containment from the bottom, a tunnel was constructed under the ruins of Block 4. Even under "normal" conditions this would have been a complicated job, and was all the more so in the intense radiation fields underneath the destroyed reactor core.

As in the case of the roof cleanup, most of the work had to be done by hand, with shovels and wheelbarrows. The protective gear and respirators hampered the workers so much that they often had to be forced to put it on.

To a large extent it was possible to overcome the technical problems and to finish the containment. But 500 million tons of radioactive debris are still stored in temporary "tombs".

And since then, the world began to learn of the first delayed effects of radiation poisoning: disturbances in hormone production and mutations. Pictured are malformed leaves and needles of trees.

Malformed children and animals are being born in ever increasing numbers.

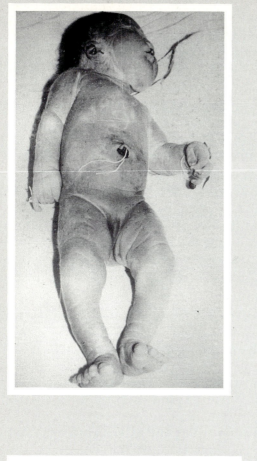

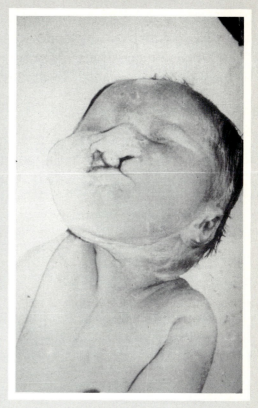

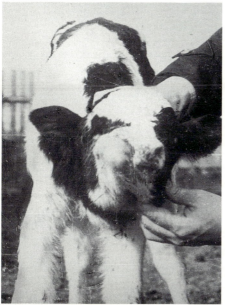

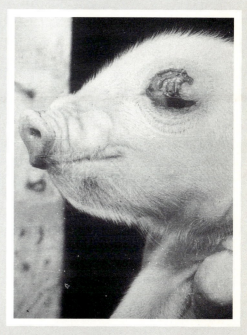

In consultations with doctors, residents exposed to radiation ask anxious questions. "Doctor, will I live?" is one of them. "Should we evacuate?" is another.

In spite of all the alarm and anxiety, some people have returned to their homes in the polluted regions. It is not possible to continue life as it was before, but at least one can die in familiar surroundings.

This tree stood as a memorial to Soviet partisans who were hung from it during World War II. It was allowed to remain standing despite being heavily irradiated. In 1990 a storm damaged it to the degree that it could no longer be preserved.

Appendix

A glossary and a list of acronyms are followed by technical details on nuclear reactors in general, and the RBMK reactors in particular, and then by details on the way in which nuclear power plants (relative to conventional ones) affect the environment. The last part contains the tables and illustrations with scientific data referred to in the various chapters of the main text.

Appendix A. Glossary and Acronyms

In this section the basic terminology pertaining to ionizing radiation and the nomenclature used to express radiation exposure and illness is briefly defined. Then a list of the (English) abbreviations is given.

Several parameters are used to quantify ionizing radiation.

— *Activity* A: the number of decay events dN occurring in a population of N_0 atoms in a time interval dt, $A = dN/dt = \lambda N = 0.693N/T_{1/2}$. The coefficient λ is the decay constant (the probability of an atom decaying in a unit time) and $T_{1/2}$ is the half-life.

— *volume* and *surface activities* are the activities divided by the dimensions of the volume or surface element, respectively. The unit of measurement is the Bequerel, 1 Bq = 1 decay event per second, or the Curie, 1 Ci = $3.7 \cdot 10^{10}$ Bq.

— *Current* F: the number of particles dN passing a given point in a time interval dt, $F = dN/dt$.

— *Fluence* Φ: the number of particles impinging on a surface element dS, $\Phi = dN/dS$.

— *Current density* Ψ: the current of particles impinging on a surface element dS (for example, the circular cut through a spherical volume element) $\Psi = d/dS = d^2N/dt \cdot dS$.

— *Radiation albedo*: the ratio of the number of particles that are reflected at the boundary between two media to the total number of impinging particles.

Similarly, several parameters quantifying the amount of energy that is transferred by the radiation to a given body have been defined. The amount of energy that is transferred to a material dW is the difference between the total energy of all charged and neutral particles (neglecting their rest energies) entering a given volume element of the material, and the total energy of all the particles leaving this volume, plus the change in energy derived from the rest masses of particles formed in nuclear reactions within this volume.

— *Kerma* or *kinetic energy released in matter*, is the ratio of the total initial kinetic energies dE_k of all charged particles formed by indirect ionizing radiation in a given volume element to the mass of this element given by $K = dE_k/dm$.

— *Photon radiation exposure* X is the ratio of the increase of the total charge (of particles of the same sign) dQ in a given volume element due to collisions with ionizing radiation to the mass of the volume element, $X = dQ/dm$. This is measured in Coulombs per kilogram or Roentgen, 1 R = $2.58 \cdot 10^{-4}$ C/kg.

A.2 Radiation Dosage

— *Absorbed energy dose* D is the ratio of the absorbed radiation energy dW to the mass of the volume element dm, $D = dW/dm$. It is measured in units of Gray or rad: 1 J/kg = 1 Gy = 100 rad.

— *Coefficient of relative biological efficiency* η is the ratio of a known absorbed energy dose D_0 to the absorbed radiation D, $\eta = D_0/D$. Usually, the calibrating radiation D_0 is $180 - 250$ kV in the x-ray wavelength range.

— *Equivalent dose* H is the product of the energy dose and some quality factors:

$$H = D \cdot Q \cdot N = \sum_j D_j Q_j N_j \, ,$$

where j denotes the kind of radiation (e. g., gamma, beta, etc.), Q is a dimensionless "biological irradiation factor", and N a normalizing coefficient which is itself, in theory, the product of several other coefficients. Usually, N is set equal to 1. The values of Q are:

—   1 for x-rays, gamma rays, and electrons;

— 10 for neutrons, protons, and singly charged ions of any energy;

— 20 for alpha particles and multiply charged ions of any energy.

The equivalent radiation dose, measured in Sievert, where 1 Sv = 1 J/kg = 100 rem, is used to quantify constant low exposure levels. For humans the safe level is set to 250 mSv/year. This parameter cannot be used as a guideline in assessing health effects of sudden high exposures such as those that occur in accidents.

— *Mean equivalent dose* H_T is the average value of H in tissue with mass m_T.

— *Effective equivalent dose* $H_E = \sum w_T \cdot H_T$ is the sum of mean equivalent doses multiplied by weighting factors w_T proportional to the *radiation sensitivity* of the given tissue. These factors are derived by measuring the ratio of some effect of irradiation-induced biological damage in a given organ or tissue to the average damage observed throughout the body.

— *Collective equivalent dose* S, in units of man-Sv, is the sum of individual equivalent doses H_i for a given group of individuals: $S = \sum_i H_i P_i$, where P_i is the number of persons receiving an equivalent dose of H_i. The per-person dose is calculated as $H_E/N_P = \dot{H}_E$ or $H_T/N_P = \dot{H}_T$, where N_P is the number of individuals. The total dose received by the gonads of a population group, including children conceived after the radiation exposure, divided by the number of individuals is known as the *genetically significant dose*.

— *The power* of the exposure, the kerma, or the absorbed or equivalent doses is simply the change of the quantity in a unit time interval:

$$\dot{X} = dX/dt \ , \quad \dot{K} = dK/dt \ , \quad \dot{D} = dD/dt \ , \quad \dot{H} = dH/dt \ .$$

— *The weakening factor* k_w is the value by which the radiation current density or the power of the dose must be reduced to a safe level.

A.3 Effects of Ionizing Radiation on the Human Body

The biological effects of radiation on tissue are the sum total of morphological and functional changes in the body occurring as a result of being irradiated. The *critical organs* are those tissues, organs, or parts of the body which, when irradiated, are likely to critically influence the health of the individual or its offspring. Different reactions are evoked in *chronic* and *acute* exposures, in which the latter refers to a single, short dose which can induce sudden illness. A *fractional irradiation dose* refers to multiple exposures separated in time.

A *radiation reaction* is a reversible change in the tissue, whereas *radiation damage* involves pathological changes which can be reversible or irreversible. *Somatic effects* are those where the individual suffers radiation damage but not its offspring; *genetic effects* damage the offspring as well and induce pathological mutations of the genes. The *somatic dose* is the radiation level needed to induce somatic damage.

We define different levels of radiation reaction and damage as: *Radiation sickness* is a systemic disease with well-defined symptoms; *acute radiation sickness* results from acute radiation exposure; for humans the approximate exposure level which induces acute radiation sickness is 1 Gray. *Primary* symptoms appear within a few weeks of exposure; *delayed* radiation effects can appear after several years.

Radionuclides are classified according to their effects on the organism.

— *Genetically significant radionuclides*: radioactive nuclides which primarily affect the gonads upon inhalation or ingestion;
— *Osteotropic nuclides*: nuclides that accumulate in bone tissue;
— *Metabolizing nuclides*: nuclides that are incorporated in the usual chemical exchange and transport processes in the body.

The *absorption coefficient* is the ratio of the amount of a radioactive substance absorbed by the blood versus the total amount absorbed by the entire body. The *accumulation coefficient* is the amount of a radioactive substance absorbed by a given organ from the blood versus the total amount of radioactive substance in the blood. *Natural elimination* of nuclides occurs when no measures are taken to accelerate or slow down elimination and when there are no pathological changes which prevent the normal flow of biological exchange processes. The effective time required to eliminate one-half of the absorbed radioactivity by natural elimination processes is $T_{\text{eff}} = T_\delta T_{1/2}/(T_\delta + T_{1/2})$, where T_δ is the time required to naturally eliminate one-half of the nuclides in the absence of radioactive decay and $T_{1/2}$ is the halflife.

A.4 Radiation Safety

Radiation safety encompasses administrative, technical, medical and sanitation measures to limit and to prevent exposure to ionizing radiation. At nuclear power plants, allowed equivalent dose levels are set for internal and external radiation and for both the personnel of the plant and the inhabitants of the surrounding area. These levels are defined under normal operating conditions and also in the event of an accident.

Let us define a few terms.

— *Category A individuals*: persons working directly with sources of ionizing radiation.

— *Category B*: persons living on monitored territory, who do not work directly with sources of radiation, but because of their place of work, type of work or place of residence can suffer the effects of radioactive substances and other sources of radiation which are discharged together with other wastes.

— *Category C*: the general population.

— *Critical group*: individuals, who because of health or age are most likely to be affected by radiation.

— *Internal irradiation*: the exposure of internal organs and tissues to ionizing radiation being emitted by decaying radionuclides.

— *External irradiation*: exposure of the skin to direct ionizing radiation and decaying nuclides.

— *Allowed Equivalent Dose* (AED): the value of the annual equivalent dose which can be received for 50 years with no ill-effects on the health of Category A individuals.

— *Allowed Dose* (AD): the maximum equivalent dose per person allowed for Category B individuals. This dose does not include radiation related to work activities, medical treatments or natural background radiation.

— *Annual Absorbtion Limit* (AAL): the amount of radionuclides that can be ingested or inhaled by Category B individuals per year, such that after 70 years the equivalent dose is equal to 1 AD.

— *Allowed dose power* (allowed equivalent dose power): the ratio of the AED or AD to the time of exposure per year. In the USSR for Category A the allowed exposure time is 1700 h/year = $6.1 \cdot 10^6$ s/year. For Category B, the allowed time if 8800 h/year = $3.2 \cdot 10^7$ s/year.

— *50% Lethal Dose*: the minimum exposure at which 50% of a population dies within a given time period (30 days). For humans this is about 4 Gy.

— *Lethal Dose*: the minimum exposure at which 100% of a population dies within a given time period (usually 30 days after exposure).

— *Radiation accident*: the violation of the limits of radioactive discharge or ionizing radiation emission established for normal operating conditions. Requires interruption of normal operating procedures and shutting down of apparatus containing sources of ionizing radiation.

— *Radiation incident*: an event in which persons are exposed to radiation at higher than allowed levels.

— *Accidental irradiation*: unforeseen increased external irradiation and/or entry of radionuclides into living tissue of personnel and local inhabitants as a result of a radiation accident or incident.

— *Nuclear accident*: an accident involving the fuel elements of the nuclear reactor or the critical nuclear assembly and accidental exposure of personnel due to:
  • loss of control over the nuclear fission chain reaction in the core (the critical nuclear assembly);
  • attainment of critical mass during transfer, transport, or storage of fuel elements;
  • breakdown of the cooling system.

— *Design-based accident* (DBA): an accident which occurs under normal operating conditions. Such accidents are taken into account in emergency planning for the safety of personnel and inhabitants.

— *Maximum design-based accident*: a (M)DBA which occurs when the reactor is operating at maximum capacity.

— *Hypothetical accident*: an accident for which no technical or administrative emergency plans to protect personnel and inhabitants are made.

A.5 Acronyms

First the English abbreviation is given, then the complete expression, and finally the transcription of the Russian abbreviation in square-brackets.

AD   – allowed dose – [DD]

AED   – (maximum) allowed equivalent dose – [PDD]

ARI   – acute radiation illness – [OLB]

ARS   – acute radiation illness – [OLB]

AS   – Academy of Sciences – [AN]

AURMRC   – All-Union Radiation Medical Research Center – [VNITsRM]

BSSR   – Byelorussian Socialist Soviet Republic – [BSSR]

BUPS   – back-up protection system – [DAZ]

ChNPS   – Chernobyl Nuclear Power Station – [ChAES]

DBA   – design basis accident

ECS   – emergency cooling system

ECCS   – emergency core cooling system – [SAOR]

ES   – emergency system – [AZ]

FAP   – fast-acting accident prevention system – [BAZ]

GSG   – General Safety Guidelines for the Planning (of NPS) – [OPB(AES)]

IAE   – Institute of Atomic Energy (in Moscow) – [IAE]

IAEA   – International Atomic Energy Agency (in Vienna) – [MAGATE]

MCP   – main circulation pumps – [GTsN]

MDBA   – maximum design basis accident – [MPA]

MDAPS   – maximum design accident protection system

MFCS   – multiple forced circulation system – [KMPTs]

MFCS   – recirculation loop – [KMPTs]

MHA   – maximal hypothetical accident – [MGA]

NPS   – nuclear power station – [AES]

NRPC   – National Radiation Protection Commission

OSR   – operational surplus of reactivity – [OZR]

RBMK   – high-power channel reactor – [RBMK]

RSN   – Radiation safety norms

SM   – supplementary moderators – [DP]

SRDIET   – Scientific Research and Development Institute of Energy Technology – [NIKIET]

SRE   – senior reactor engineer

SRNPS   – Safety Regulations for Nuclear Power Stations – [PBYa]

TBSRI   – technical basis of safety of reactor installation – [TOBRU]

TG   – turbine generator – [TG]

SAR – shortened absorber rod – [USP]
VCS – ventilation and cooling system – [CPIR]

A.6 Tables of Appendix A

Table **A.1.** Units of activity and dose

Parameter	Unit	SI Unit[a]	Conversion	
Activity	Curie [Ci]	Becquerel [Bq] 1 event/second	1 Bq =	$2.7 \cdot 10^{-11}$ Ci
Energy dose	rad [rd]	Gray [Gy] 1 Gy = 1 J/kg	1 Gy =	100 rd
Equivalent dose	rem [rem]	Sievert [Sv] 1 Sv = 1 J/kg	1 Sv =	100 rem
Ion dose	Roentgen [R]	Coulomb per kilogram	1 C/kg ≈	3876 R

[a] Système Internationale

Table **A.2.** Medical effects that occur in the event of short-term exposures, as would be the case in a nuclear power plant accident or bomb test.

≤ 50 rem	Very small doses cannot be checked by blood tests and the like; the closer they approach the 50 rem the easier it is to show slight changes in the blood.
≤ 120 rem	5 to 10% of patients experience vomiting, nausea and fatigue for the first 24 hours.
≤ 170 rem	25% of patients experience vomiting, nausea, and fatigue for the first 24 hours, followed by other symptoms. Expected probability of death is zero.
≤ 260 rem	25% of patients experience vomiting, nausea, and fatigue for the first 24 hours, followed by other symptoms. Expected probability of death is greater than zero.
≤ 330 rem	Nearly all patients experience vomiting and nausea in the first 24 hours, followed by other symptoms, 20% of patients die within 2 to 6 weeks, survivors require 3 months convalescence.
≤ 500 rem	All patients suffer vomiting and nausea in the first 24 hours, followed by other symptoms, 50% of patients die within 1 month, survivors require 6 months convalescence.
≤ 770 rem	All patients experience vomiting and nausea within 4 hours, followed by other symptoms. Probability of death is nearly 100%. The few survivors require at least 6 months convalescence.
≤ 1000 rem	All patients experience vomiting and nausea within 1 to 2 hours. No survivors.
≤ 5000 rem	Immediate vomiting. Death within 1 week.

Appendix B. Reactor Technology

B.1 The Fueling Reaction

The energy source in a nuclear reactor is the chain reaction of fission of heavy nuclei induced by collisions with neutrons. Nuclides that are fissile and can participate in a chain reaction are ^{233}U, ^{235}U, ^{239}Pu, ^{241}Pu and other transuranium elements. Uranium, the most commonly used fuel, in its naturally occurring state is a mixture of 0.006% ^{233}U, 0.7% ^{235}U and 99.3% ^{238}U [A.1]. The fuel is enriched by isotope separation techniques, converted to UO_2 powder and baked into pellets.

In the interaction of neutrons with the nuclei of U-238, the absorption of a neutron without subsequent fission is much more probable than the fission reaction. Therefore, U-238 acts as a neutron absorber which hinders the chain reaction. For this reason, the naturally occurring uranium can be enriched with U-235, by more than an order of magnitude. This is done for *fast neutron reactors*.

The reaction of U-235 is about 1000 times *more* efficient with thermal energy neutrons than with fast neutrons. Thus, in *thermal neutron reactors*, naturally occurring (or slightly enriched, by 1.8% to 4.4%) uranium fuel is used along with some kind of neutron decelerator (moderator) elements.

In every reaction a neutron is absorbed and two atomic fragments, several neutrons, and other particles are produced. The total energy released in one reaction event is ~ 200 MeV; 5 MeV are distributed among the product neutrons. This energy is millions of times that obtained in the combustion of organic fuels.

Most of the neutrons produced are "prompt" neutrons, formed within 10^{-12} s after the nuclei are split. About 0.65% of the product neutrons are "delayed" by a second or more.

Part of the product neutrons continue the chain, producing the next generation of neutrons, and part are either absorbed without inducing further fission or escape into the surrounding medium. Thus, in order for the chain reaction to be maintained, the number of neutrons produced must be greater than or equal to the total number of neutrons that are absorbed, lost, and required to continue the chain. The ratio of product neutrons to reactant neutrons is called the *neutron multiplication coefficient* k. If $k = 1$, the chain reaction is maintained. If $k > 1$ then the rate of the reaction increases, and if $k < 1$ the reaction is, or will eventually be, quenched.

The relative difference of the coefficient k from unity $\varrho = (k - 1)/k$ is called the *parameter of reactor reactivity*. In the critical state ϱ is equal to zero; in the supercritical and subcritical states it is greater than and less than zero, respectively. The reactor reactivity is also measured in units of β or the number of manual fuel rods in the core.

The reactor power is controlled by the number of neutrons. The number of neutrons produced with each successive generation will increase until $k = 1$, which thus yields the maximum operating power. Obviously, to lower the power it is necessary to let k drop below 1.

The change in the power $Q(t)$ as a function of time is $Q(t) = Q_0 \exp \varrho t/\theta$, where Q_0 is the power at $t = 0$ and θ the mean lifetime of a neutron generation. The time required for the reactor power to change by a factor of e (2.718) is called the *reactor period*. Commonly, the *power doubling period* $T_{1/2}$ is also used, which is related to the reactor period by $T_{1/2} = 0.693T$. If $\varrho = 0.005$ and $\theta = 10^{-3}$ s, then in 1 second the power will increase by a factor of 150.

Controlling such extremely rapid processes is exceptionally difficult. But it is not impossible. If the delayed neutrons are taken into account in calculating the mean lifetime of a neutron generation, then θ is about 0.1 sec, allowing the reactor power to increase only by a factor of 1.05 rather than 150 for $\varrho = 0.005$.

If the reactor reactivity is greater than or equal to the 0.65%-fraction of delayed neutrons, then only the prompt neutrons are needed in order for the chain reaction to be sustained. Such a reactor is said to be in a prompt critical state. It undergoes a rapid exponential increase in power and can easily go out of control. For this reason, *the positive reactor reactivity is always kept at less than 0.0065*.

During reactor operation, the fuel is burned and fragments which might quench the neutrons are accumulated. This lowers the reactivity parameter. Therefore, for prolonged periods of operation, with several power-up and power-down cycles, it is necessary to have an *initial reactivity*. This initial re-activity is obtained by loading enough fuel to exceed the *critical mass*, which is the mass of fuel in a *critical volume* at which the neutron escape rate is equal to their rate of formation.

After the fuel is burned down to the point of inefficient power production, the remainder is removed for reprocessing. The length of time that a reactor can operate on a single fuel charge, the *reactor campaign*, is not only determined by the density and volume of the fuel, but also by the durability of the fuel rod holders, which are exposed to high pressures of gaseous fission products, neutron radiation, and very high temperatures.

The *fuel burnup* is the possible energy yield per ton of reactor fuel. This is primarily dependent on the fuel enrichment: the higher the fraction of fissile isotopes, the higher the burnup. The burnup is also dependent on the type of fuel pellet (metal or ceramic), its porosity, and the material and thickness of the fuel element cladding.

In most reactors, for naturally occurring uranium, one can attain 1000 – 3000 MW·day/ton, for low-grade enriched uranium 3500 – 10 000; intermediate enrichment gives 10 000 – 50 000, and superenriched uranium fuels yield up to 100 000 MW·day/ton.

Note that the reactor power is determined by the rate of the fission reaction in the core. Since one reaction releases 200 MeV of energy, to obtain 1 W of power it is necessary to have $3.15 \cdot 10^{10}$ fission events per second. Maintaining a power of 1 MW for 24 hours thus requires 1.225 g of uranium-235. A typical 3000 MW reactor consumes 4 kg/day of U-235.

The Core: To maintain a steady chain reaction, the masses of the fuel and the dimensions of the reactor volume must be chosen appropriately. The rate of neutron loss is dependent on the physical and chemical properties of the materials and also on their shape, since neutrons are produced throughout the reactive volume but lost only through its surface. The nuclear fuel is contained in *fuel rods* (elements) in the core. To compensate for the excess initial reactivity, usually neutron absorbing rods made of boron, cadmium, hafnium, or rare-earth elements are introduced.

The heat emitted by the fuel elements is carried away by the heat carrier (coolant) which is continuously circulated through the core.

Usually, the reactor core is surrounded by a layer of neutron reflecting material, i.e., a material which has a low probability of neutron absorption and high scattering probability. Because the neutron moderator has to meet the same requirements, frequently the moderator and reflector are made of the same substance.

The core, the reflector, and other reactor elements are placed inside a *biological shield.*

Reaction Control and Protection: The *Reactor Control and Protection System* (RCPS) initiates the reaction, maintains it, and shuts it down (also for emergency shutdowns). In addition to the fuel element rods and moderator rods, special (emergency) rods with very high neutron absorbing quotients made of boron, cadmium, or other materials are included in this system.

The insertion of these rods into the RCPS channels of the core stops the fission reaction. Initiation and regulation is achieved by partial or complete extraction.

B.2 Reactor Types

Thermal Neutron Reactors: Thermal neutron reactors, as mentioned above, are distinguished by the presence of a neutron moderator. This moderator can be contained in special elements in the core or it can be the heat carrier (coolant) used to power the turbines. The moderator must have an atomic mass low enough so that upon collision with the neutrons a sufficient amount of energy can be transferred. It must also have a low neutron absorption coefficient and high activation threshold. The most commonly used moderator substances are water, heavy water, and graphite.

Fast Neutron Reactors: The main feature which distinguishes fast neutron reactors from other types is the *blanket* surrounding the core. It is filled with a convertible heavy element (such as natural or depleted uranium or thorium) which absorbs the escaping neutrons.

Breeder Reactors: An important property of nuclear reactors is their ability to produce synthetic fissile nuclides. For example, capture of neutrons by U-238 forms plutonium-239, and naturally occurring thorium-232 is converted to U-233. Thus, along with energy production, there is, in principle, the possibility of converting the stable ^{238}U and ^{232}Th nuclides to nuclear fuels. This process is parameterized by a conversion coefficient which is equal to the ratio of the number of nuclides produced to the number of nuclei consumed.

Under certain conditions the production of the unstable plutonium-239 and uranium-233 can exceed the amount of fuel that is consumed, that is, the coefficient is greater than one. Such reactors are called breeder reactors. This property is used to produce other types of reactor fuels, for example, plutonium mixed with natural or low-grade enriched uranium, plutonium with thorium, U-233 with thorium, etc..

B.3 Series Reactors

In the USSR, as well as in other countries, most nuclear power plant reactors follow two designs [A.2 – 5] using thermal neutrons and low-grade or naturally occurring uranium: In water-water reactors (WWR, or WWER) water is used both as a neutron moderator and as a heat carrier (coolant), and in high power facilities (HPCR for High Power Channel Reactor or RBMK — the Russian abbreviation has been adopted in the West) the reactor is composed of graphite-moderated and water-cooled channels.

Water-Graphite Channel Reactors: In this reactor design, the core is placed in a concrete shaft (for an RBMK-1000 this shaft is $21.6 \times 21.6 \times 25.5$ m^3) and is held in place by a metal cradle several meter above the floor of the shaft. The core is in a graphite box which is cooled by a helium-nitrogen gas flow. The fuel (uranium pellets) and cooling (water containing) element rods are arranged in a chessboard pattern and enter through the top of the graphite box. The RBMK design allows fuel replacement without interrupting reactor operation.

Heat generated in the fuel rods is carried away from the reactor core by the water, which is heated to the boiling point. The steam is separated from the water in a separator drum and is fed to a turbine which powers a generator. After the steam cools, the condensed water is pumped back into the separator drum.

Characteristics of power plants of the RBMK–1000 and RBMK–1500 series are given in Tables B.1 and B.2. There is no reactor of a design analogous to this RBMK used outside of the USSR. In certain aspects (e. g., single-circuit operation, boiling heat carrier), the RBMK is similar to reactors using water as a heat carrier and moderator, that is, boiling water and pressurized water reactors (BWR, PWR).

In RBMK and WWR reactors the sealed metallic (zirconium alloy) claddings of the fuel elements prevent leakage of radioactive fission products from the fuel into the coolant. Due to corrosion, however, these claddings develop defects (usually microcracks) through which light particles can escape. Together with the corrosion products, these particles contaminate the water and make it radioactive. Therefore, a bypass system is installed to filter the sediments and to regulate the water acidity (specifically, the boric acid concentration).

In single-circuit RBMKs, the steam used to drive the turbine can become radioactive due to neutron activation of the oxygen, as well as from the above contributions. Therefore, stringent radiation safety measures are required in this area of the power plant.

B.4 Tables of Appendix B

Table **B.1.** Parameters of WWR and RBMK power plants

Reactor type:	WWR-440	WWR-1000	RBMK-1000	RBMK-1500
Power [MW$_{el}$]	440	1000	1000	1500
Net power efficiency [%]	32	33	31.3	31.3
First-circuit pressure [MPa]	12.3	15.7		
Heat carrier temperature [°C]				
at reactor inlet	269	289		
at reactor outlet	300	322	280	280
Mass of first-circuit heat carrier [tons]	200	300	–	. –
Uranium charge [tons]	42	66	192	189
Mean initial uranium enrichment [%]	3.5	3.3	1.8	1.8
Uranium consumption [tons/yr]	14	33	50	75
Mean degree of burnout [MW-day/kg U]	28	30	18.1	18.1
Reactor dimensions				
height (without dome) [m]	11.8	10.88	–	–
maximum diameter [m]	4.27	4.57	–	–
mass [tons]	200.8	300	–	–
Core dimensions				
height [m]	2.5	3.55	7.0	7.0
diameter [m]	2.88	3.1	11.8	11.8

Table B.2: Some technical data for the RBMK reactor are presented.

Effect of reactivity	Magnitude			Typical rise time
	%	β	Number of manual regulator rods	
1 2	3	4	5	6
1 Depoisoning Xe and Sm	3.58	7.16	72	48 h
2 Firing up (20–270 °C)	2.13	4.26	42	25 h
3 Steam void	2.25	4.5	45	3 sec after
	4.05	8.1	81	neutron power
	7.44	14.88	149	jump
	3.0	6.0	60	
4 Fuel Doppler	−0.55	−1.1	−11	2 sec after
	−3.04	−6.08	−61	neutron power jump
5 Replacement of He by N_2	0.55	1.1	11	8–10 h
6 RCPS Evacuation	2.5	5.0	50	−15 sec (ES only)
7 Recharging 1/14 of all fuel channels	2.55	5.1	51	several days
8 Full power	1.3	2.6	26	5 h
9 Attainment of positive reactivity using RCPS (within safety regulation specifications)	0.15	0.3	4	17–24 sec
10 Attainment of positive reactivity using RCPS in the lower part of the reactor(1259 mm in height) for $t > 11$ sec, excess reactivity ~ 0, ES-5 activated	1.0	2.0	20	3 sec after ES-5 switch
	3.65	7.3	73	activation
11 Operational excess reactivity	3.75	6.5	65	17–24 sec
	6.25	12.5	125	
12 Attainment of negative reactivity using the ES-5	All RCPS rods without excess reactivity and shortened absorber rods			17–24 sec

Appendix C. Nuclear Power Plants as Sources of Environmental Pollution

C.1 Operational (Regular) Emissions

The large-scale adoption of any new technology cannot be decided for based only on its economic or technological merits. It must also be judged on its ecological integrity, based on the risk to which it exposes society and the environment. Experience has shown thus far that the ability of an ecosystem to regenerate itself is not unlimited. Although significant work is being done in developing low-waste and zero-waste technologies, at present they are not useful in power-plant design. In an on-line power plant, gaseous, liquid, and solid wastes are produced. Of these, aerosols, and to a lesser degree, liquids, are the primary pollutants.

Four barrier systems are placed in the path of escaping radioactive species:

1) *The fuel matrix.* The high melting point (2760°C) and chemical stability of uranium oxide hinders the escape of fission products except in extreme conditions. If the core does not melt, then the UO_2 pellets trap 98% of the accumulated radionuclides.
2) *The fuel rod cladding.* This cladding is made of a zirconium-niobium alloy that also confines the nuclides to the fuel rod.
3) *The cooling water system.* The water lines are pipes with 20-cm thick walls, surrounded by multiple layers of steel. The system is continuously checked for leaks and cracks.
4) *Reactor shielding.* This shielding is usually a concrete dome of 1.2 m thickness, steel plated and armored with steel rods.

Thus, the discharge of radioactive species is only possible when all four barriers are overcome.

The sources of radioactive contamination are the products of neutron activation produced in the fuel rods and fission products which leak from the fuel rod into the heat carrier. As mentioned above, WWR and RBMK reactors are efficient breeders of synthetic radioactive particles formed in the fission of U-233, U-235, and Pu-239, and in the neutron activation of the materials constituting the core of the reactor.

The total radioactivity of these species is dominated by that of the short-lived nuclides, and it reaches 10^{10} Ci for the entire core. The specific radioactivity of the mixture of the fission products attains $2 \cdot 10^6$ Ci per ton of fuel. After 3 years, the specific radioactivity decreases to $7 \cdot 10^5$ Ci/ton, and the absolute activity is $3 \cdot 10^7$ Ci for the core [C.1].

Table C.1 shows that 98% of the radioactive particles are produced in the fuel rods and remain in them until the heat flux in the core is sufficiently high to induce leakage through the rod cladding. This source of contamination is weak for all nuclides except tritium. The diffusion of tritium through zirconium cladding is 1% , and through steel claddings it is 80% [C.1]. Table C.2 shows the radioactivity of the fission-produced nuclides which accumulate in a 3500 MW (thermal power) reactor over the course of a 1-year operation period. Increase of the period to 5 years does not affect the radioactivity due to short-lived species, but it does increase the activity coming from strontium-90 and cesium-137 by nearly a factor of five.

Other radionuclides are further classified roughly according to their chemical properties which determine their potential toxicity in the environment. *Air-borne* materials are of two types: *Inert* materials are the noble gases (rare gases) such as krypton and xenon. These gases are easily transported through the reactor system and into the environment. *Chemically active* substances such as halogens (iodine, bromine) are very water soluble, they form acids, and have a low vaporization temperature; alkali metals (cesium, rubidium) react easily with water to form hydroxide bases (e.g., CsOH) or salts with the dissolved halogens (CsI); tellurides (tellurium, selenium, antimony), are not as mobile as halogens and alkalis, and they decay to iodine; ruthenium forms an oxide that can be air-borne over quite long distances.

Non-air-borne species are the alkali earths (barium, strontium), the noble metals (ruthenium, rhodium, palladium), the rare–earths (yttrium, lanthanum, cerium, praseodymium, neodymium, promethium, neptunium, plutonium) and refractory oxides of zirconium, niobium, molybdenum, and technetium.

In general, in the event of an accident, all radionuclides produced in the core can be released into the environment. However, their relative amounts will differ due to fractionation – normally, more gases will be released (under atmospheric conditions) than solid contaminants. The non-air-borne contaminants may enter the surrounding environment when the integrity of the cooling of the fuel elements is violated and the temperature approaches thousands of degrees.

Aerosols: The primary sources of aerosol wastes is the bypass system of the first circuit (in WWR or PWR plants) and the ejector of the condenser (in RBMK or BWR plants). Degassing through leaks in the cooling water circuit, evaporation during water recycling, and water sampling are contributory sources.

Before being discharged into the atmosphere, these gases undergo a complex treatment to remove the iodine and other aerosols. They are temporarily trapped to allow the short-lived radionuclides to decay. The gases are then pumped into holding tanks or are passed through a "radiochromatographer" for purification.

The final discharges are typically composed of rare gases, tritium, and the remaining iodine and aerosols. The rare gas fraction consists of krypton and xenon and the neutron activation product ^{41}Ar. Iodine atoms are bound in organic molecules (90%) with an admixture of aerosolic and elemental iodine

(10%) also present. The half-lives of these species are listed in Table C.3. In Table C.4 are listed the various sources of tritium produced in WWR and RBMK reactors. Some fission and activation products which are in the solid state under normal conditions are also found in the gaseous waste [C.1]. Table C.5 summarizes the gas phase emissions from Soviet WWR and RBMK power reactors according to nuclide; Table C.6 again shows the main components for several Soviet nuclear power stations.

The primary difference between RBMK and WWR facilities is that the second has a closed first water circuit. In these reactors the radioactive nuclides spend a much longer time within the reactor than in the open first circuit of the RBMK. The water flux out from the first circuit of a WWR to the bypass filter system or through leakage is relatively small. Thus, the time it takes for half of the radioactive gas to leave the first circuit is long — from a few hours to several days, depending on the type of reactor and its operating regime [C.1].

In a boiling RBMK gases are transferred from the water to the steam. They rather rapidly (within 30 minutes) leave the circuit through the condenser ejector. Therefore, in atomic power plants of this type, the temporary confinement of gases before discharge through a stack is more important than in reactors of WWR type [C.1]. The amount of rare gases in RBMK power plants is thus an order of magnitude larger than in WWR facilities. In gaseous wastes of WWRs the primary component is the relatively long-lived ^{133}Xe, while in RBMK wastes the major fraction is composed of the shorter lived krypton and xenon isotopes (Table C.7).

The gas circuit of the RBMK acts as a secondary source of ^{3}H, ^{14}C, and ^{41}Ar. Purification and temporary confinement of the gases from this circuit allow effective containment of the carbon and argon, so that the carbon discharge is reduced to the same level as that from WWR power plants. The tritium produced in the circuit is emitted together with other gaseous wastes. The amount of ^{14}C entering the atmosphere is not more than $6 - 10$ mCi/MW(electrical)-year (this is two times lower than from a radionuclide-producing chemical plant); iodine-129 discharges are negligible.

The absolute discharge levels of rare gases from WWR power plants are 10^3 Ci/year; those of aerosols of long-lived species are a few tens of mCi/year or less. The content of ^{131}I in gas-aerosol emissions is also quite low (several tens of mCi/year or less). For power plants with channel reactors the rare gas emissions are about $10^4 - 10^5$ Ci/year, while aerosols and iodine amount to less than 10 Ci/year [C.2].

Liquid Radioactive Wastes: The liquid radioactive wastes of nuclear power plants include cooling water, wash waters used to clean the facility, as well as laundry and sewage waters, etc.. All these waters are purified and deactivated and then recycled. The amounts and compositions of nuclides in liquid wastes vary significantly with reactor type and power capacity. Isotopes typically found in waste waters (as well as in solid wastes) are listed in Tables C.8 (products of fission reactions) and C.9 (products of neutron activation).

Under normal reactor operation the radioactivity of the cooling water (without tritium) is 10^{-4} Ci; with the appearance of defects in the fuel rod cladding this rises to 10^{-2} Ci. The maximum detected activity of the coolant is on the order of 1 Ci [C.3]. Table C.10 compares the amounts of liquid waste fractions produced in three types of reactors. Table C.11 shows the averaged data of relative and absolute amounts of radionuclides in waste water.

The specific radioactivity of the liquid waste fraction is $3 \cdot 10^6 - 1.2 \cdot 10^8$ Bq/l and is primarily due to ^{137}Cs (50 – 90%), ^{134}Cs and ^{51}Co (10 – 35%). The fraction of the remaining nuclides is less than 5%. The concentration of tritium is an order of magnitude larger in WWR power plants than in RBMK plants and is $10^4 - 10^5$ Bq/l (less than 10%). The ^{90}Sr fraction is not higher than 0.1% [C.1].

During reactor operation some quantity of surplus wash waters (trapping, deactivating, and rinse waters of the laundries) are produced which, after their radioactivity levels are measured, are transferred to cooling reservoirs. Under normal operating conditions the radionuclide discharge into these reservoirs is not high: 5 and 1 TBq [GW(elec.)-yr] of tritium for WWRs and RBMKs, respectively; for other nuclides this is five to seven orders of magnitude lower [C.4].

The ^{137}Cs and ^{134}Cs content in liquid wastes of this type is 30 – 50% in WWRs and RBMKs, accounting for 70% of the total radioactivity. The ^{131}I fraction is 10 – 40%, ^{58}Co, ^{60}Co is about 15% of WWR liquid wastes [C.4]. It should be emphasized that the scatter in these data is quite high and is strongly dependent on the reactor type [C.5]: for boiling water reactors we have at Monticello $2.9 \cdot 10^{-6}$ Ci/year at Mills Town 51.5 Ci/year, and in high-pressure reactors, at Yankee Row 0.206 Ci/year, and San Onofre 30.3 Ci/year.

The specific activity of the deionized waste waters transferred into the cooling reservoirs is usually low, less that $2 \cdot 10^{-10}$ Ci/l, and is not greater than the concentration allowed in drinking water ($3 \cdot 10^{-10}$ Ci/l). This radioactivity is primarily due to tritium and to small amounts of other nuclides (^{134}Cs, ^{137}Cs, ^{60}Co, etc.; Table C.12) which are not removed in the decontamination [C.1].

The water treatment system does not allow the tritium isotope to escape. Thus, this nuclide is the major contaminant in the reservoirs, especially in WWR plants. The absolute value of the tritium concentration in the first circuit is, in BWRs, $1.5 \cdot 10^{-5}$ Ci/l, and in PWRs, $1.5 \cdot 10^{-3}$ Ci/l [C.6]. The volume of the first circuit in WWRs and RBMKs is 400 m^3 and 1200 m^3, respectively [C.3]. This yields a radioactivity of 2 Ci/MW-yr for WWRs and for RBMKs 0.1 Ci/Mw-yr [C.7]. Without the tritium fraction, the total radioactivity in the first circuit is 0.001 Ci/MW-yr for a WWR and 0.01 Ci/MW-yr for an RBMK.

That is, in comparison to tritium, the amount of radioactivity due to fission and activation products is small, reaching only a few or tens of Curies for 1000 MW per year. The total radioactivity of nuclides (excluding tritium) collected in the water-cooling tanks amounts to 1 Ci/year for WWR and 10 Ci/year for RBMKs [C.2].

Thus, the amount of radioactive substances leaked into the environment by liquid wastes is not significant if one excludes tritium, and it is less than the aerosol emission. The amount of tritium put out by WWR plants is high, however: 1500 – 2000 Ci/year [C.2].

Emission of liquid radioactive wastes also contaminates the bottoms of the water reservoirs. Even at low contamination levels ($< 2 \cdot 10^{-10}$ Ci/l), sedimentation results in accumulation of radionuclides, especially in finely dispersed silt. These sediments in RBMK cooling reservoirs contain ^{60}Co, ^{51}Cr, ^{54}Mn, ^{65}Zn, ^{134}Cs and ^{137}Cs, even at significant distances (3 km) from the point of discharge [C.7]. Analysis of data taken over several years on discharges into the reservoirs of large WWRs in the Federal Republic of Germany (the former West Germany) showed that the discharge of ^{137}Cs and ^{134}Cs accounts for 5 and 3.1 mCi/MW(elec.)/year, and that of ^{89}Sr and ^{90}Sr is 0.63 and 0.063 mCi/MW(elec.)/year [C.7].

Power plants that produce no radionuclidic discharges into reservoirs do exist. For example, the deionized waters of the Novovoronezh nuclear power plant are fed to filtration fields where the radioactive nuclides are completely trapped in the surface soil layer [C.8].

At present, systematic measurements of the concentration of radioactive substances in rivers and lakes in the vicinity of nuclear power stations do not allow one to discern any influence of the stations on the radioactivity of structures in the external environment — it does not rise above background levels.

Solid Wastes: Solid wastes can arise from several sources such as solidified liquid wastes, whose radioactivity is primarily caused by the ^{106}Ru, ^{95}Zr, ^{98}Nb, activated corrosion products; scrapped equipment and used ventilation filters, construction materials, polishing materials, protective clothing, etc.. A large part of the solid waste is usually of low contamination levels. However, although intermediate and highly radioactive wastes account for only several percent of the volume, they constitute the major contribution to the radioactivity. They contain practically all the radioactive nuclides extracted during purification of the gaseous and liquid wastes.

The normalized volumes of the solid wastes of varying specific radioactivity, calculated on the basis of observational data on the Soviet reactors WWR-440, RBMK-1000 and BN-350, are given in Table C.13. The values show that 70 – 80% of the total volume is due to low-level waste, 40 to 60% of which is organic material.

The allowed amount of radioactivity of solid wastes of light water reactors (including solidified deposits and ion-exchanger gels) is 200 Ci/(MW year) [C.9, 10]. Solid wastes are temporarily stored at the power station and then transferred to special storage facilities.

C.2 A Comparison: Radioactive Contamination from Combustion of Fossil Fuels

Naturally occurring nuclides are present in all geological formations, including fossil fuels such as coal, peat, and oil. Of these, the solid fuels, such as rock coal, are probably the more important source of nuclides, because they are the most widely used in energy production. The major contaminating element is ^{226}Ra. Its concentration varies over a broad range, usually between 0.3 and 2.9 pCi/g (0.011 – 0.1 Bq/g) in various types of coal. Table C.14 shows the content of naturally occurring nuclides of the more common coal types.

Irradiation of humans occurs in mining, which puts the mine personnel at risk, and during combustion, when the general populace is exposed to emissions. In agreement with many studies, in coal mines the ^{222}Rn gas content and that of its daughter isotopes varies over two and even three orders of magnitude [C.1]. This mean concentration is about an order of magnitude lower than in uranium mines, i. e., 10^{-11} Ci/l.

The cumulative dose to which miners are exposed in a year is approximately 7 Sv/person. This dose is about four times greater than the total radiation dose that uranium miners receive. This seems contradictory at first. However, although the concentration of nuclides in the air of coal shafts is about 10 times lower than in the air of uranium mines, the amount of usable coal (4 million tons) is 40 times larger than the amount of usable uranium (10^5 tons), which does lead to the obtained result.

During fuel combustion ^{226}Ra is adsorbed on solid particles and can enter the environment together with air-borne ash. The redesorbing nuclides are incorporated into biological food chains and thus enter the human body through foods, inhaled air, and drinking water.

A 1000 W thermoelectrical power station annually consumes 2 Ci ($7.4 \cdot 10^{10}$ Bq) of ^{226}Ra during coal combustion, of which about 10 mCi (10^8 Bq) are put into the atmosphere together with ash. Facilities using rock coal are thus major sources of radioactive air pollution, accounting for 2.5 – 28% of the ^{226}Ra in the lower atmospheric layers [C.1].

The area of the region contaminated by aerosols from thermoelectrical power plants is strongly dependent on a number of factors: the discharge flux, the effectiveness of the waste-purification system, the height of the smoke stack, meteorological conditions, etc.. The radionuclide composition of air-borne ash can also be quite varied. Several authors have reported ^{226}Ra concentrations between 0.4 and 6.4 pCi/g (0.015 – 0.237 Bq/g). The effective equivalent radiation dosage due to coal-fired power plants is significantly higher than that of nuclear power plants of similar capacity. This is true even when the effectiveness of ash entrapment by smoke filtration systems is assumed to be 98.5%, which is a reasonable value for some modern coal plants. The level of radiation due to naturally occurring nuclides to which a population in the vicinity of a nuclear power plant (with an RBMK) is exposed is ~5 times less than the population near a coal-fired plant. For facilities using WWR-440 reactors this number is about 40.

The radiation doses concur with recent estimates made in the Federal Republic of Germany, the United States, and Canada, which saw a 2 to 10 times higher dosage level for coal-fired than for nuclear power plants.

C.3 Accidental Emissions

Nuclear energy has the highest potential for danger among all the energy sources presently known. Therefore, in the construction of nuclear power plants, particular attention must be given to safety procedures in crisis situations, which are impossible to prevent completely [C.11].

The greatest danger in a reactor accident is not due to the large number of radioactive particles accumulated in the core, but rather the emission of these fission products into the environment from a damaged reactor. Unfortunately, it is very difficult to foresee the amount of accidental emission since it depends on a number of unpredictable factors. The discharge of radionuclides can vary over 3 to 6 orders of magnitude under the same hypothetical initial causes and extent of reactor damage, depending on the course of the accident, the chemical properties of the released substances, whether they are gaseous or solid, their temperature, etc.. Despite the unavoidable inaccuracy, such prognoses are quite important in evaluating the potential radiation spill and in planning for accident prevention and containment.

Three types of accident categories have been defined: local, regional, and extended. In the first type, the discharge of pollutants is confined to a single building or apparatus. In a regional accident, the emitted nuclides are distributed over the site of the plant, whereas in an extended event the contamination spreads beyond the facility.

In Table C.15 the measured levels of accidentally discharged nuclides at various power plants are compiled. These data show that all of these have been local events except for those at Chernobyl and Windscale, which surpass even the optimistic expectations of the "extended" category, and are really better described as "global".

The advantage of having water in the containment shell to localize the discharge, not only within the limits of a power plant, but also beyond them is shown by these data. For example, in the Windscale accident a discharge volume of 10 m^3 of highly radioactive substances (total activity of 10^5 Ci) contaminated only 700 m^3 of soil to a level of 1 Roentgen per hour [C.12]. The 1979 Three Mile Island accident demonstrated the reliability of multiple shields in confining radionuclides emitted from the core, which kept this event to a local level [C.13].

More significant discharges occur in accidents in which the reactor is damaged. Not only are radioactive rare gases ejected into the atmosphere, but nuclides which are volatile at the melting point of the nuclear fuel are also ejected. The result of such an accident is a significant contamination of large areas and the subsequent long-term withdrawal of the land from productive use. The Chernobyl disaster eliminated more than 3000 km^2.

C.4 Tables of Appendix C

Table C.1. Radioactivity levels of nuclides in a typical 1 GW (elec.) nuclear power plant

	Total radioactivity [Ci]			Fraction of the activity in the core [%]		
	Fuel	Container Walls	Total	Fuel	Container Walls	Total
Core^a	8.1 · 10^9	1.4 · 10^8	8.24 · 10^9	98.0	2.0	100
Spent fuel storage						
maximum^b	1.3 · 10^8	1.3 · 10^7	1.31 · 10^9	16.0	0.16	16
average^c	3.6 · 10^8	3.8 · 10^6	3.64 · 10^8	4.5	4.8 · 10^{-2}	4.5
Transport container^d	2.2 · 10^7	3.1 · 10^5	2.23 · 10^7	0.27	3.8 · 10^{-3}	0.27
Transfer container^e	2.2 · 10^7	2 · 10^5	2.2 · 10^7	0.27	2.5 · 10^3	0.27
Waste storage facility						
gaseous	–	–	9.3 · 10^4	–	–	1.2 · 10^{-3}
liquid	–	–	9.5	–	–	1.2 · 10^{-6}

^{a} Calculated for a thermal power of 3.2 GW, 30 min after shutdown
^{b} With two-thirds of the core unloaded: one-third 3 days previously and one-third 150 days previously
^{c} With one-half of the core unloaded: one-quarter stored 150 days, one-quarter stored 60 days
^{d} Contains 7 PWR or 17 BWR cooling elements after 60 day operation stop
^{e} Contains one cooling element after 3-day operation stop

Table C.2. Radioactivity of nuclides in the reactor core
(3500 MW thermal power)

Group/radionuclide	Activity [MCi]	Half-life [days]
Tritium	0.025	4507
Noble gases		
^{55}Kr	0.60	3950
^{55}Kr	26	0.183
^{57}Kr	51	0.0528
^{58}Kr	73	0.117
^{133}Xe	183	5.28
^{135}Xe	37	0.384
Iodine		
^{131}I	91	8.05
^{132}I	129	0.0958
^{133}I	183	0.0875
^{134}I	204	0.0366
^{135}I	161	0.280
Alkali metals		
^{86}Rb	0.028	18.7
^{134}Cs	8.1	750
^{136}Cs	3.2	130
^{137}Cs	5.1	11000
Tellurium-antimony		
^{127}Te	6.3	0.391
^{127}Te	7.2	109
^{129}Te	33	0.048
^{129}Te	5.7	34
^{131}Te	14	1.25
^{132}Te	129	3.25
^{127}Sb	6.6	3.88
^{129}Sb	35	0.179
Alkali-earth metals		
^{89}Sr	101	52.1
^{90}Sr	4.0	11030
^{91}Sr	118	0.403
^{140}Ba	172	12.8
Noble metals and cobalt		
^{58}Co	0.84	71
^{60}Co	0.31	1920
^{103}Ru	118	39.5
^{105}Ru	77	0.185
^{106}Ru	27	366
^{105}Ru	53	1.5

Group/radionuclide	Activity [MCi]	Half-life [days]
Rare-earths, refractory oxides, transuranium elements		
^{90}Y	4.2	2.67
^{91}Y	129	59
^{95}Zr	161	65.2
^{97}Zr	161	35
^{95}Nb	144.6	35
^{99}Mo	172	2.8
^{99}Tc	151	0.25
^{140}La	172	1.67
^{141}Ce	161	32.3
^{143}Ce	140	1.38
^{144}Ce	91	284
^{143}Pr	140	13.7
^{147}Nd	65	11.1
^{239}Np	1800	2.35
^{238}Pu	0.061	32500
^{239}Pu	0.023	$8.9 \cdot 10^6$
^{240}Pu	0.023	$2.4 \cdot 10^6$
^{241}Pu	3.7	5350
^{241}Am	0.0018	$1.5 \cdot 10^5$
^{242}Cm	0.54	163
^{244}Cm	0.025	6630

Table C.3. Radionuclides of noble gases and iodine formed in nuclear reactors. These species emit β or γ-radiation

Nuclide	$T_{1/2}$	Nuclide	$T_{1/2}$	Nuclide	$T_{1/2}$
^{85}Kr	10.7 yrs	^{133}Xe	5.2 days	^{129}I	$1.6 \cdot 10^7$ yrs
^{85}Kr	4.5 h	^{134}Xe	2.2 days	^{131}I	8 days
^{87}Fr	1.3 h	^{135}Xe	9.1 h	^{133}I	21 h
^{88}Kr	2.8 h	^{136}Xe	15.7 min	^{135}I	6.6 h

Table C.4. Calculated amounts of tritium produced in a power reactor.

Reaction	Ci/MW (elec.)-yr	
	WWR	RBMK
Fission	18–20	18–20
Neutron activation		
^{2}H	0.001	0.01
^{6}Li	0.02	0.5 [a]
^{10}Be	0.5	–
^{3}He	–	< 0.6 [b]

[a] In lithium-doped graphite
[b] In the RBMK gas flow system

Table C.5. Radioactivity of gaseous wastes of nuclear power plants [Bq/GW (elec.)-yr]

Nuclide	WWR	RBMK
Rare gases	$1.6 \cdot 10^{14}$	$4.7 \cdot 10^{15}$
^{3}H	$4.4 \cdot 10^{12}$	$1.9 \cdot 10^{12}$
^{14}C	$3.0 \cdot 10^{11}$	$1.4 \cdot 10^{12}$
^{89,90}Sr	$6.2 \cdot 10^{7}$	$7.7 \cdot 10^{7}$
^{131}I	$1.3 \cdot 10^{10}$	$3.6 \cdot 10^{10}$
Mixture of long-lived nuclides [a]	$7.2 \cdot 10^{9}$	$1.7 \cdot 10^{10}$

[a] species with $T_{1/2} > 24$ h, exept ^{3}H, ^{14}C, ^{89,90}Sr, ^{131}I

Table C.6. Gas-aerosol discharges

Nuclear power plant	Discharge radioactivity [%]			Total allowed gaseous discharges Ci/MW(elec.)-yr
	Rare gases	Aerosols	^{131}I	
Novovoronezh	2.1	4.1	1.4	4.3
Kolsk	1.1	0.13	0.02	3.0
Armenia	1.2	1.3	4.3	6.0
Chernobyl	36.0	20.0	4.0	82.0
Kursk	18.6	6.7	0.9	80.0

Table C.7. Radioactive rare gas and iodine content in gaseous emissions of nuclear power plants

Nuclide	Fraction [%]	
	WWR	RBMK ^{a}
^{41}Ar	0.2	0.3
^{85}Kr	6.0	0.7
^{86}Kr	5.4	6.6
^{87}Kr	1.0	13.4
^{88}Kr	2.2	18.6
^{133}Xe	72.0	35.2
^{135}Xe	13.2	25.4
Total	100	100
^{131}I	59.8	23.8
^{133}I	31.9	43.5
^{135}I	8.3	32.7
Total	100	100

^{a} Without ^{41}Ar from the gas circuit

Table C.8. Solid waste products of fission reactions

Nuclide	$T_{1/2}$	Nuclide	$T_{1/2}$	Nuclide	$T_{1/2}$	Nuclide	$T_{1/2}$
^{89}Sr	51 days	^{95}Zr	64 days	^{134}Cs	2.1 yrs	^{143}Pr	14 days
^{90}Sr	28.6 yrs	^{103}Ru	39 days	^{137}Cs	30 yrs	^{144}Ce	284 days
^{91}Y	59 days	^{106}Ru	1 yr	^{140}Ba	13 days	^{155}Eu	5 yrs
^{95}Nb	35 days	^{129}Te	34 days	^{141}Ce	33 days		

Table C.9. Products of neutron activation

Nuclide	$T_{1/2}$	Nuclide	$T_{1/2}$	Nuclide	$T_{1/2}$	Nuclide	$T_{1/2}$
^{51}Cr	28 days	^{59}Fe	45 days	^{95}Nb	35 days	^{3}H	12.3 yrs
^{54}Mn	312 days	^{60}Co	5.3 yrs	^{95}Zr	64 days	^{14}C	5730 yrs
^{58}Co	71 days	^{65}Zn	244 days	^{110}Am	250 days	^{41}Ar	1.8 h

Table C.10. Volumes of radioactive wastes of nuclear power plants discharged with liquids [10^3m^3/GW (elec.)-yr]

Material	WWR	RBMK	BN
Total:			
with realistic salt content	1.0–4.6	0.7	0.6
with projected salt content [400 g/l]	0.4–1.7	0.35	0.06
Ion-exchange gels	0.01–0.04	0.04	–
Activated charcoal	0.002–0.0025	0.003	–
Perlite	–	0.15–0.20	–

Table C.11. Radioactivity of the first-circuit waters in units of maximum allowed doses

Nuclide	Important daughter nuclide	Radioactivity	
		PWR	BWR
^{3}H	–	10^3	10
^{24}Na	–	10–200	110
^{54}Mn	–	5	10
^{58}Co	–	10	1–100
^{60}Co	–	2	0.2–400
^{63}Ni	–	10	–
^{64}Cu	–	10	10–30
^{89}Sr	–	3	30
^{90}Sr	^{90}Y	1	30
^{91}Sr	^{91}Y	1–5	5–350
^{93}Y	–	0.03	200
^{95}Zr	^{95}Nb	1	150
^{95}Nb	–	0.5	5.0
^{99}Mo	^{99}Tc	1	10
^{103}Ru	^{103}Rn	1	10
^{106}Ru	^{106}Rn	0.1	10
^{110}Ag	^{110}Ag	1	0.003
^{122}Sb	–	10	–
^{131}I	–	100–300	100–30000
^{132}I	^{132}I	0.1	25
^{133}I	–	800–1000	200–60000
^{135}I	–	10	100000
^{134}Cs	–	3	0.01–1
^{137}Cs	^{137}Ba	1	1
^{140}Ba	^{140}La	1	0.02–50
^{141}Ce	–	0.1	10
^{143}Ce	^{143}Pm	0.1	100
^{144}Ce	^{144}Pm	0.1	30
^{147}Pm	^{147}Pm	–	50
^{152}Ta	–	1	0.5
^{187}W	–	5	3
^{239}Np	^{239}Pu	0.01	200

Table C.12. Annual radionuclide discharge in liquid wastes of nuclear power plants [Ci/yr]

Nuclide	BWR		PWR	
	Minimum	Maximum	Minimum	Maximum
^{3}H	$7.6 \cdot 10^{-5}$	$1.2 \cdot 10^2$	5.03	$5.89 \cdot 10^3$
^{14}C	–	–	$1.71 \cdot 10^{-2}$	0.33
^{24}Na	$3.08 \cdot 10^{-2}$	0.141	$1.58 \cdot 10^{-2}$	1.32
^{51}Cr	$5.6 \cdot 10^{-2}$	1.3	$2.2 \cdot 10^{-4}$	2.77
^{54}Mn	$1.07 \cdot 10^{-3}$	4.67	$5.08 \cdot 10^{-4}$	2.36
^{59}Fe	0.02	0.109	0.067	0.139
^{58}Co	$7.9 \cdot 10^{-7}$	27.1	$2.67 \cdot 10^{-4}$	3.17
^{60}Co	$8.58 \cdot 10^{-4}$	2.83	2.0	–
^{65}Zn	$2.3 \cdot 10^{-2}$	0.355	$3.9 \cdot 10^{-3}$	–
^{89}Sr	$7.8 \cdot 10^{-4}$	1.03	$4.15 \cdot 10^{-5}$	0.02
^{90}Sr	$7.8 \cdot 10^{-4}$	0.216	$4.66 \cdot 10^{-6}$	$1.29 \cdot 10^{-3}$
^{103}Ru	–	–	0.032	0.3
^{131}I	0.052	11.4	$1.29 \cdot 10^{-3}$	1.95
^{134}Cs	0.237	8.86	$1.64 \cdot 10^{-4}$	8.66
^{137}Cs	$8.3 \cdot 10^{-7}$	16.4	$2.33 \cdot 10^{-4}$	6.44
^{140}Ba	0.05	0.367	$1.46 \cdot 10^{-4}$	0.041
^{144}Ce	0.025	–	$2.3 \cdot 10^{-5}$	0.227
^{239}Np	0.04	0.683	–	–

Table C.13. Solid radioactive waste volume of nuclear power plants [m^3/GW (elec.)-yr]

Specific activity	[Sv/h] at 0.1 m above ground level	Specific β-activity [Bq/kg]	WWR	RBMK	BN
High	> 10	$3.7 \cdot 10^9$	20	10	20
Medium	0.3–10	$3.7 \cdot 10^6 - 3.7 \cdot 10^9$	170	150	150
Low	$3 \cdot 10^{-4} - 0.3$	$7.4 \cdot 10^4 - 3.7 \cdot 10^6$	660	500	600
Total	–	–	850	660	770

Table C.14. Radionuclide content in coal and coal combustion products

	^{238}U	^{226}Ra	^{210}Pb	^{232}Tn	^{228}Tn	^{228}Ra	^{40}K
Coal							
USSR	–	0.5–2.9	0.2–1.4	0.3–2.0	–	–	–
Poland	–	0.001–1.3	–	–	–	–	–
Czechoslovakia	–	0.7	0.6	–	–	0.11–0.35	25
USA	0.49	0.3	–	0.57	–	–	1.4
Coal ash							
USSR	–	5.0	–	2.3–8.7	–	3.3	–
USA	–	3.8	–	2.6	–	2.4	–
Slag							
Poland	–	4.3	–	–	1.2	–	–
USA	4.9	0.55–4.5	1.0	1.5	0.5	–	26
Flying ash							
Poland	0.6–6.7	1.0–6.4	–	0.18–0.2	1.0	–	22.5
USA	10	0.4–3.1	17.3	2.6	0.4	–	11.5

Table C.15. Reactor accidents

Facility; Thermal power capacity [MW]	Radioactive discharge [Ci]			
	Iodine	Fission products		Rare gases
		Nuclide	Activity	
Windscale 250	$2 \cdot 10^4$ (12%) in atmosphere	Tellurium Caesium-137 Strontium-89 Strontium-90 in atmosphere	1600 600 80 9	$3.4 \cdot 10^5$ in atmosphere
SL-1 (Idaho)	80 (0.5%) in atmosphere	Strontium-90 Caesium-137	0.1 0.5 surface	$1 \cdot 10^4$ in atmosphere
NRX National Research Experimental Reactor Chalk River, Canada 30.0	no data	total	$1 \cdot 10^4$ [a] H_2O	no data 28
TMI-2 Three Mile Island 2720	17	none in atmosphere		$10 \cdot 10^7$ in atmosphere
WTP Westinghouse Testing Reactor Waltz Mill Pennsylvania, USA	none	total	10^4 [b]	800 in atmosphere
CR-3 Crystal River, Florida	70 [c]	no data		10^3 in atmosphere
Plutonium Recycle Test Reactor Hanford, Washington	205 (27%) [d]	no data		50 in atmosphere
HTRE-3 Heat Transfer Reactor Experiment Arco, Idaho USA 0.12	34 (14%) in atmosphere	Strontium-90 total	0.1 400 in atmosphere	no data
ORR Oak Ridge Laboratory	0.2 in atmosphere	total	1000 [e]	no data
Chernobyl 3200	$7 \cdot 10^6$	total	$50 \cdot 10^6$-$6.4 \cdot 10^9$	$180 \cdot 10^6$

[a] In $4 \cdot 10^6$ l cooling water of the hermetic shielding
[b] In $7 \cdot 10^6$ l cooling water of the hermetic shielding
[c] In $1.6 \cdot 10^5$ l cooling water of the hermetic shielding; 2.0 in air
[d] In the cooling water; 7.0 in air
[e] In the first-circuit cooling water

Appendix D. Data Related to Chapters 10 and 11

The tables and graphic representations given here are, in part, explicitly referred to in those chapters, they partially supplement discussions and information presented in those chapters.

D.1 Data Supplementing Chapter 10

Table 10.5. Settlements where radiation levels are being regulated by lowering the cesium-137 content in milk (by as much as a factor of 2).

No.	Settlement	Population	Total dose until 1990	Radiation dose [rem] over 70 years		
				external	internal	total
Bryanskaya Oblast'						
Gordeyevkii Rayon						
1.	Mirnyi	2123	6.38	9.28	28.19	43.85
2.	Gordeyevka	2165	6.60	9.60	19.10	45.30
3.	Bezboshnik (now Bol'shevik)	41	5.26	7.65	23.56	36.47
4.	Krishtopov Rychei	34	6.12	8.90	27.10	42.12
5.	Staro-Novitskoe	318	5.57	8.10	24.83	38.50
6.	Shiryaevka	400	5.90	8.58	26.19	40.67
7.	Palomy Sukryn	9	6.16	8.96	27.29	42.41
8.	Pokrovka	7	5.19	7.55	23.29	36.03
9.	Rogovets	2	5.79	8.42	25.74	39.95
Klitsovskii Rayon						
10.	Drobnitsa	27	5.39	7.84	24.10	37.33
11.	Krasnyi Luch	49	5.68	8.26	25.29	39.23
12.	Uletovka	63	6.58	9.57	29.01	45.16
13.	Chakhov	51	6.18	8.99	27.38	42.55
14.	Neprino	273	6.58	9.57	29.01	45.16
15.	Gorelaya Sosna	10	5.35	7.78	23.93	37.06
16.	Kamenka	2	5.21	7.58	23.38	36.17
Krasnogorskii Rayon						
17.	Krylovka	3	6.07	8.83	26.92	41.82
18.	Baturovka	244	5.65	8.22	25.20	39.07
19.	Stobunka	27	5.24	7.62	23.47	36.33
20.	Novo-Mikhailovka	7	5.04	7.33	22.65	36.02
21.	Beryozovka	32	7.28	10.59	31.92	49.79
22.	Mikhalevka	23	6.53	9.50	28.83	44.86
23.	Novo-Drozhinsk	70	5.50	8.00	24.56	38.06
24.	Novaya Zhizn'	3	5.26	7.65	23.56	36.47
25.	Chigrai	17	6.53	9.50	28.83	44.86
26.	Zaozer'yev	33	5.13	7.46	23.02	35.61
27.	Lesnoi	44	5.76	8.38	25.65	39.79
28.	Uvel'ye	701	6.67	9.70	29.38	45.75
29.	Ust'ye	2	6.29	9.15	27.83	43.27

Novozybkovskii Rayon

30.	Zlynka	5800	6.38	9.28	28.19	43.85
31.	Rassadniki	62	5.94	8.64	26.38	40.96
32.	Groznyi	82	5.30	7.71	23.74	36.75
33.	Krasnaya Zaya	55	6.12	8.90	27.10	42.12
34.	Borshchovka	40	6.53	9.50	28.83	44.86
35.	Klyukov Mokh	28	5.61	8.16	25.02	38.79
36.	Pavlovka	8	6.18	8.99	27.38	42.55
37.	Gremuchka	6	6.07	8.83	26.92	41.82
38.	Kalinin	1	6.82	9.92	30.01	46.75
39.	Sviderki	11	5.15	7.49	23.11	35.75
40.	Savichka	16	5.72	8.32	25.47	39.51
41.	Rudnya	34	6.53	9.50	28.83	44.86
42.	Krivoi Sad	37	6.84	9.95	30.10	46.89
43.	Demenka	399	6.38	9.28	28.19	43.85
44.	Perevoz	194	6.34	9.22	28.01	43.57
45.	Barki	393	5.70	8.29	25.38	39.37
46.	Sennoye	129	6.01	8.74	26.65	41.40
47.	Muravinka	58	5.92	8.61	26.29	40.82
48.	Krasnyye Orly	20	6.64	9.66	29.28	45.58
49.	Lyubin	137	5.52	8.03	24.65	38.20
50.	Vikholka	329	5.24	7.62	23.47	36.33
51.	Zhuravka	166	5.08	7.39	22.83	35.30
52.	Podles'ye	1	5.52	8.03	24.65	38.20
53.	Pobeda	23	6.62	9.63	29.19	45.44
54.	Novyye Bobochi	962	6.07	8.83	26.29	41.82
55.	Selinkoye	20	5.21	7.58	23.38	36.17
56.	Makusy	4	5.83	8.48	25.92	40.23
57.	Makusy-2	1	5.90	8.58	26.19	40.67
58.	Granitsa	32	7.26	10.56	31.82	49.64
59.	Novoye Mesto	611	5.72	8.32	25.47	39.51
60.	Shplomy	752	5.06	7.36	22.74	35.16
61.	Karny	89	6.38	9.28	28.19	43.85
62.	Novaya Derevnya	3	7.06	10.27	31.01	48.34
63.	Palomy	9	5.37	7.81	24.02	37.20
64.	Yasnaya Polyana	61	5.21	7.58	23.38	36.17
65.	Grivka	27	5.92	8.61	26.29	40.82
66.	Staryye Bobovichi	1224	5.92	8.61	26.29	40.82
67.	Kurgan'ye	36	6.80	9.89	29.92	46.61
68.	Prudovka	45	7.11	10.34	31.19	48.64
69.	Kolodezskii	9	6.27	9.12	27.74	43.13
70.	Khalivichi	542	6.67	9.70	29.38	45.75
71.	Mashkinskii	38	5.72	8.32	25.47	39.51
72.	Yagodnoye	86	7.13	10.37	31.28	48.78

Kievskaya Oblast'
Polesskii Rayon

73.	Rudnya Grezlyanskaya	148	8.68	8.96	27.29	44.93
74.	Polesskoye	11800	8.37	8.64	26.38	43.39

Chernobyl'skii Rayon

75.	Zimovishche	0	8.37	8.64	26.38	43.39

Zhitomirskaya Oblast'
Narodichskii Rayon

76.	Zvezdal'	305	6.79	7.01	21.74	35.54
77.	Sloboda	105	6.82	7.04	21.84	35.70
78.	Velikiye Kleshchi	550	9.27	9.57	29.01	47.85

Mogilevskaya Oblast'
Klimovichskii Rayon

79.	Selishche	16	8.40	12.80	17.86	39.06
80.	Budishche	27	6.84	8.96	28.88	44.68
81.	Kanchary	16	7.83	10.56	27.19	45.58

Kostyukovichskii Rayon

82.	Kletki	131	9.97	11.52	21.05	42.54
83.	Dubiyets	94	6.48	8.64	22.86	37.98
84.	Gaikova	73	9.45	14.40	12.02	35.87
85.	Voronovka	97	8.35	11.20	19.16	38.71
86.	Prudok	56	5.45	7.36	28.27	41.08

Krasnopol'skii Rayon

87.	Goslev	20	8.87	11.84	23.19	43.90
88.	Yel'nya	117	4.62	7.04	28.52	40.18
89.	Radilev	66	7.90	10.56	27.84	46.30
90.	Dragotyn'	100	7.26	9.92	21.68	38.86
91.	Berezyaki	148	4.83	7.36	22.87	35.06
92.	Vydrenka	368	18.70	5.12	33.18	57.00

Cherikovskii Rayon

93.	Veprin	928	6.93	10.56	17.61	35.10
94.	Dubrovka	81	4.83	7.36	28.03	40.22
95.	Ostrovy	3	10.21	13.76	14.23	38.20

Gomel'skaya Oblast'
Vuda-Koshelevskii Rayon

96.	Luk'yanovo	21	6.30	9.60	29.10	45.00

Vetkovskii Rayon

97.	Leski	62	9.03	13.76	18.88	41.67
98.	Komylin	20	5.46	8.32	25.47	39.25
99.	Pervomaisk	59	6.51	9.92	39.01	46.44
100.	Kositskoye	166	5.67	8.64	22.29	36.60
101.	Krasnyi Ugol	68	6.51	9.92	21.72	38.15
102.	Staroye Zakruzh'ye	568	5.04	7.68	23.65	36.37
103.	Amel'noye	81	5.25	8.00	24.56	37.81
104.	Novoivanovka	25	6.93	10.56	31.82	49.31
105.	Gutka	89	8.19	12.48	25.41	46.08

Dobrushskii Rayon

106.	Bol'shoi Les	84	5.04	7.68	24.56	37.28
107.	Selishche-2	168	3.78	5.76	34.21	43.75
108.	Berezki	236	6.51	9.92	30.01	46.44
109.	Morozovka	284	5.25	8.00	24.56	37.81
110.	Yasnaya Polyana	6	5.46	8.32	25.47	39.25

Kalinkovichskii Rayon

111.	Nekrashevka	247	3.36	5.12	38.75	47.23
112.	Kuz'michi	275	3.57	5.44	44.99	54.00

Narovlyanskii Rayon

113.	Khomonki	103	5.52	7.36	22.74	35.62
114.	Konotop	491	6.72	8.96	20.02	35.70
115.	Smolegovskaya Rudnya	201	6.48	8.64	31.94	47.06
116.	Luben'	175	6.48	8.64	26.38	41.50
117.	Zarakitnoye	137	6.24	8.32	25.47	40.03

Khoinikskii Rayon

118.	Omel'kovshchina	302	8.40	9.90	25.70	43.70
119.	Markhlevsk	176	6.72	7.68	23.65	38.05

Checherskii Rayon

120.	Rudnya Dudicheskaya	112	6.51	9.92	30.01	46.44

Table 10.6. Settlements where agricultural improvements and mechanical soil decontamination is necessary to lower the cesium-137 level in milk by a factor of 2.

No. Settlement	Population	Total dose until 1990	Radiation dose [rem] over 70 years		
			internal	external	total
Bryanskaya Oblast'					
Klintsovskii Rayon					
1. Kuznets	206	8.58	12.48	37.27	58.33
Novozybkovskii Rayon					
2. Filial "VIUA"	415	7.88	11.46	34.37	53.71
3. Krasnyi Kamen'	66	7.52	10.94	32.92	51.38
4. Medvezh'ye	56	7.83	11.39	34.19	53.41
5. Borshch	28	7.74	11.26	33.82	52.82
6. Bilimovka	8	7.63	11.10	33.37	52.10
7. Moshok	10	8.40	12.22	36.55	57.17
8. Staryi Vyshkov	897	7.74	11.26	33.82	52.82
Zhitomirskaya Oblast'					
Narodichskii Rayon					
9. Peremoga	65	13.79	14.24	22.74	50.77
10. Khristinovka	368	9.61	9.92	30.01	49.54
Ovruchskii Rayon					
11. Gladkovicheskaya Kamenka	68	10.54	10.88	32.74	54.16
Mogilevskaya Oblast'					
Klimovichskii Rayon					
12. Aleksandrovka	32	9.87	15.04	9.23	34.14
Kostyukovichskii Rayon					
13. Derazhnya	387	12.97	17.28	17.72	47.97
14. Vetukhna	574	8.97	14.40	20.87	44.84
15. Leninskii	3	10.75	14.40	21.19	46.34
16. Mokroye	195	7.50	11.52	20.50	39.52
17. Khotimsk	244	9.73	12.80	23.90	46.43
Krasnopol'skii Rayon					
18. Gorezna	127	9.33	11.84	19.02	40.19
19. Zhelizh'ye	92	11.56	17.28	19.52	48.36
20. Manuily	66	9.69	17.28	9.19	36.16
Slavgorodskii Rayon					
21. Zapolyan'ye	35	12.98	17.28	14.79	45.05
22. Kulikovka 1	259	8.49	11.52	26.33	46.34
23. Kulikovka 2	221	8.81	13.44	10.67	32.93
Cherikovskii Rayon					
24. Novaya Malinovka	73	11.22	20.16	17.86	49.24
25. Malinovka	364	10.45	19.84	17.41	47.70
Gomel'skaya Oblast'					
Braghinskii Rayon					
26. Vyazok	301	11.47	11.83	22.63	45.94

Dobrushskii Rayon
27. Krasnyi Log   27   10.29   15.68   13.77   39.74

Kormyanskii Rayon
28. Kostyukovka   75   12.81   19.52   7.42   39.75

Checherskii Rayon
29. Serbovichi   293   8.82   13.44   23.99   46.25
30. Novoye Zarech'ye   73   9.66   14.72   7.81   32.19
31. Dubrovka   70   11.34   17.28   5.83   34.45

Total: 5698

Table 10.7. Settlements where cesium-137 levels in milk must by lowered by a factor of 4 (through agricultural improvements and the performance of mechanical soil decontamination). Entries in the table marked by an asterisk refer to places that need further radiation level measurements to test the possibility of minimising the external radiation dose.

No. Settlement	Population	Total dose until 1990	Radiation dose [rem] over 70 years		
			internal	external	total
Bryanskaya Oblast'					
Krasnogorskii Rayon					
1. Dolgoye	37	10.49	15.26	45.18	70.93
2. Aleksandrovka	80	9.68	14.08	41.82	65.58
3. Yalovka	1396	11.48	16.70	49.27	77.45
Novozybkovskii Rayon					
4. Vyshkov	3900	8.36	12.16	36.37	56.89
5. Orel	6	9.09	13.22	39.36	61.67
6. Stolpino	29	9.02	13.12	39.09	61.23
7. Savitskii Log	162	8.87	12.90	38.46	60.23
8. Chekhov	59	9.04	13.15	39.18	61.37
9. Zarech'ye	23	8.80	12.80	38.19	59.79
10. San'kovo	50	10.23	14.88	44.09	69.20
11. Kamen'	203	9.28	13.50	40.18	62.96
12. Glybochka	29	8.80	12.80	38.19	59.79
13. Borok	20	9.70	14.11	41.91	65.72
14. Svyatsk	641	10.31	15.04	44.54	69.92
15. Griva (Grivka)	48	8.56	12.45	37.19	58.20
Gordeyevskii Rayon					
16. Kozhany	939	9.99	14.53	43.09	67.61
Kievskaya Oblast'					
Polesskii Rayon					
17. Novaya Markovka	164	13.02	13.44	47.27	73.73
18. Zhavtnevoye	369	11.16	11.52	34.55	57.23
Zhitomirskaya Oblast'					
Narodichskii Rayon					
19. Nozdrishche	485	11.78	12.16	34.78	58.72
20. Khriplya	58	9.61	9.92	55.78	75.31
21. Gudnya Ososhnya	235	13.95	14.40	40.09	68.44
22. Slavenshchina	185	10.85	11.20	33.64	55.69
23. Staroye Sharno	480	12.09	12.48	37.27	61.84
24. Malyye Kleshchi	425	13.64	14.08	41.82	69.54
25. Repishche	51	23.41	34.56	32.90	90.87
Cherikovskii Rayon					
26. Osovets	5	25.14	20.48	17.86	63.48
27. Chudyany	323	24.92	45.12	36.97	107.01
Gomel'skaya Oblast'					
Vetkovskii Rayon					
28. Bartolomeyevka	762	7.77	11.84	85.87	105.48

29. Potesy	73	6.09	8.28	101.19	116.56
30. Ryslavl'	20	10.50	16.00	47.27	73.77

Dobrushskii Rayon

31. Pennoye	27	15.12	23.04	67.25	105.41

Khoinikskii Rayon

32. Lomachi	73	18.48	21.12	64.30	103.90

Checherskii Rayon

33. Shepetovichskiye Poplavy	42	14.91	22.72	73.95	111.58
34. Podosov'ye *	71	5.46	8.32	59.76	73.54
35. Budishche *	339	3.36	5.12	44.70	55.18
36. Osinovka *	208	5.04	7.68	79.05	91.77
Total:	5557				

Table 10.8. Settlements where it is not possible to achieve the allowed radiation levels (through intensive agricultural improvements and soil deactivation). Entries marked by an asterisk refer to places that need further radiation level measurements to test the possibility of minimising the external radiation dose.

No. Settlement	Population	Total dose until 1990	Radiation dose [rem] over 70 years		
			internal	external	total
Bryanskaya Oblast'					
Krasnogorskii Rayon					
1. Borki	38	16.39	23.84	69.52	109.75
2. Tugani	60	14.74	21.44	62.71	98.89
3. Gushchi	33	13.68	19.90	58.35	91.93
4. Bukuven	188	16.68	24.26	70.70	111.64
5. Kovali	54	15.18	22.08	64.52	101.78
6. Prokhorenko	34	15.51	22.56	65.88	103.95
7. Zabor'ye	953	15.58	22.66	66.16	104.40
8. Novoakelsandrovka *	59	11.53	16.77	49.45	77.75
9. Nikolaevka *	408	15.86	23.07	67.34	106.27
10. Yamnishche	77	14.01	20.42	59.80	94.26
Novozybkovskii Rayon					
11. Babaki	13	13.13	19.10	56.08	88.31
Kievskaya Oblast'					
Polesskii Rayon					
12. Shevchenko	182	15.19	15.68	46.36	77.23
13. Yasen	96	24.18	24.96	71.70	121.84
Zhitomirskaya Oblast'					
Narodichskii Rayon					
14. Polesskoye *	120	18.60	19.20	75.65	113.45
15. Shishelivka	124	16.06	16.58	48.90	81.54
16. Malyye Min'ki	179	20.15	20.80	60.89	101.84
Mogilevskakya Oblast'					
Kostyukovichskii Rayon					
17. Chervonyi Ugol	17	14.49	22.08	42.47	79.04
18. Virovka	76	15.61	18.56	47.59	81.76
Krasnopol'skii Rayon					
19. Losinka	1	12.27	16.61	51.19	80.10
20. Dragomilovo	41	12.68	19.52	80.97	113.17
21. Gotovets	166	14.67	21.12	101.53	143.32
22. Mar'yina Buda *	64	9.62	16.00	66.30	91.92
23. Novaya Yel'nya *	481	15.97	29.44	46.74	92.15
24. Bol'shoi Osov *	99	11.48	13.76	64.57	89.81
Mogilevskaya Oblast'					
Kostyukovskii Rayon					
25. Zarech'ye	60	12.36	17.28	24.02	53.68
26. Ozerets	25	10.59	14.08	30.36	55.03
27. Gutka	55	11.26	15.04	28.51	54.81
28. Mamonovka	20	8.00	10.88	34.15	53.03

Krasnopol'skii Rayon

29.	Vysokii Borok	322	11.78	21.76	15.52	49.06
30.	Mkhinichi	227	10.18	17.60	48.67	76.45
31.	Krasnaya Zarya	39	10.58	14.08	40.45	65.11
32.	Korma Dolgaya	61	8.10	14.40	54.18	76.68
33.	Malyi Osov	96	12.12	15.04	29.75	56.91
34.	Korma Paiki	234	10.88	20.88	43.43	75.11
35.	Gorodok	19	8.41	14.72	38.30	61.43
36.	Borovaya	26	7.48	12.16	32.37	52.01
37.	Zavodok	349	13.78	24.32	31.21	69.31
38.	Dubrovka	62	16.12	22.40	25.86	64.38
39.	Staraya Buda	114	9.45	11.20	38.25	58.90

Cherikovskii Rayon

40.	Kamenka	78	12.01	22.72	19.63	54.36
41.	Ushaki	340	6.96	8.96	35.27	51.07
42.	Lysovka	1	9.42	12.48	45.90	67.80
43.	Bakunovichi	265	7.16	9.60	36.35	53.11
44.	Monastyrsk	72	9.40	12.80	27.65	49.85

Gomel'skaya Oblast'
Buda-Kosheyelvskii Rayon

45.	Boyevoi	36	8.82	13.44	40.00	62.26

Vetkovskii Rayon

46.	Besed'	181	7.56	11.52	39.32	58.40
47.	Vorob'yevka	72	9.66	14.72	37.05	61.43
48.	Osovo	9	9.66	14.72	34.21	58.59
49.	Selitskoye	98	5.25	8.00	34.21	47.46
50.	Novyye Gromyki	289	6.72	10.24	51.81	68.77
51.	Rudnya Guleo	120	8.19	12.48	37.27	57.94
52.	Sivenka	463	6.72	10.24	50.67	67.63
53.	Krasnyi Pet'	2	7.35	11.20	23.64	52.19
54.	Podkamen'ye	44	8.82	13.44	40.00	62.26
55.	Khizy	123	8.19	12.48	37.27	57.94

Dobrushskii Rayon

56.	Bilevo	203	7.14	10.88	32.51	50.53
57.	Dem'yanki	972	4.83	7.36	29.10	41.29
58.	Krasnoye Znamya	43	7.14	10.88	32.74	50.76
59.	Dubetskoye	192	5.04	7.68	41.59	54.31
60.	Ploskoye	31	7.35	11.20	26.74	45.29

Checherskii Rayon

61.	Shepetovichi	194	12.60	19.20	19.06	50.86
62.	Voskhod	12	7.14	10.88	32.74	60.76

| Total: | 15575 | | | | |

D.2 Data Supplementing Chapter 11

The following data are included in this book although they were received too late to be discussed in detail. (They have been composed by M. Ankudovich from Minsk, Oncomorphological Department Minsk Patalogo-Anatomic Center, and by the respective regional executive authorities of the Health Organizations [oblispolkom], respectively.) The data allow to assess the development of some ailments during the periods 1980 – 1989 and 1986 – 1989.

Figures D.2.1 – 4 refer to data from the Vetka area in Gomel province. They give the *sickness rates per thousand children under 14 years*. The respective ailments are those of the abdominal and digestive tract, urinal, skin, and cordial diseases (Fig. D.2.1); tonsillitis and lung diseases (Fig. D.2.2); liability to anemia and clinical degrees of anemia (Fig. D.2.3); 1-st degree thyroid gland diseases (Fig. D.2.4); for 2-nd degree there are no data available for 1988 and 1989, their number for 1986 is 196 and the one for 1987 is 218; for 3-rd degree the data for the 4 years from 1986 – 1989 are 5, 1, 5, and 2, respectively.

Figures D.2.5 – 8 refer to the development of leucosis and other hemoblastoses in various provinces of Byelorussia. In Fig. D.2.5 the evolution of these ailments in the six provinces of Byelorussia is displayed. The next illustration repeats the data for the province of Minsk and compares them to the ones of the city of Minsk. The time evolution of the total numbers is shown in Fig. D.2.7. The numbers for 1989 include in all cases only the ailments of part of the year (the first 9 months!). The number of cases of leucosis and other hemoblastoses in Minsk province per 100 thousand of the *children's* population is shown in Fig. D.2.8.

Fig. D.2.1. Vetka area
Gomel province

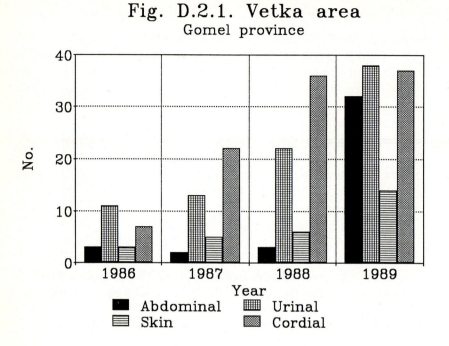

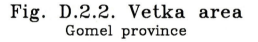

Fig. D.2.2. Vetka area
Gomel province

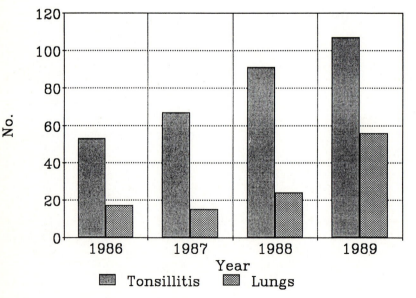

Fig. D.2.3. Vetka area
Gomel province

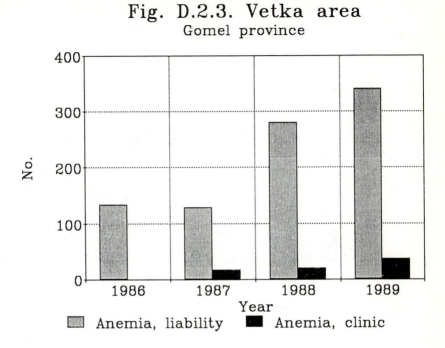

Fig. D.2.4. Vetka area
Gomel province

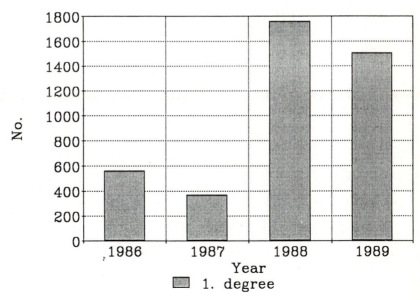

D.2.5. Leucosis
and other hemoblastoses

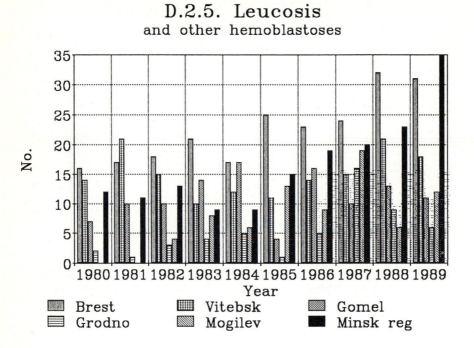

Brest   Vitebsk   Gomel
Grodno   Mogilev   Minsk reg

D.2.6. Leucosis
and other hemoblastoses

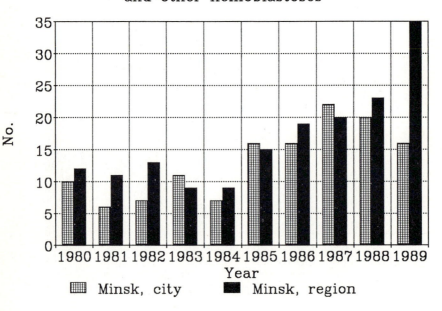

Minsk, city   Minsk, region

Fig. D.2.7. Leucosis
and other hemoblastoses

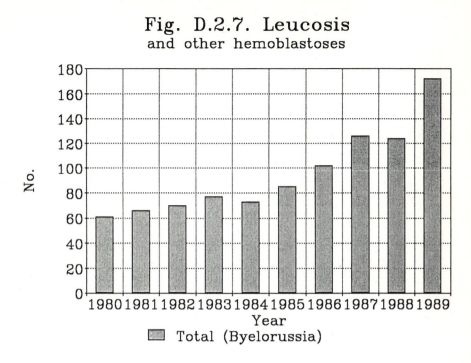

Total (Byelorussia)

Fig. D.2.8. Leucosis
and other hemoblastoses

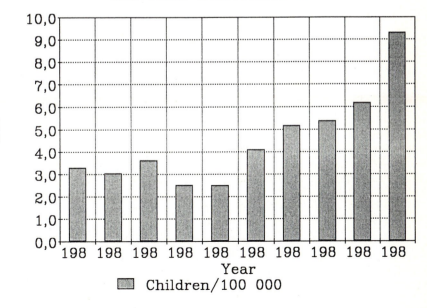

Children/100 000

References

Chapter 3

3.1 *Analis rezultatov perevoda reaktorov RBMK-1000 na toplivo s obogashce-niem 2.4 %* (IAE-Report No. 33R/1-388-89, 1989)

3.2 V.A. Khalimonchuk, A.V. Krayushkin, A.V. Kuchin, et al.: *Nejtronno-fizicheskie raschety maksimalnoj proektnoj avarii v RBMK-1000* (joint report of the Kurchatov IAE and the IAE of the AN USSR, No. 33/769987, 1987)

3.3 A.V. Krayushkin, V.A. Khalimonchuk, V.S. Romanenko et al.: *Nejtronno-fizicheskie issledovaniya MPA v reaktorakh RBMK* (joint report of the Kurchatov IAE and the INI of the AN USSR, No. 33/859588, 1988)

3.4 IAE-Report, No. 33/806587 DSP, 1987)

3.5 *Fizichekij pusk reaktora RBMK-1000 3-go bloka ChAES* (ChNPS-Report No. 229/P-3 PTO ChNPS 1981)

3.6 *Protokol rezultatov izmerenij fizicheskikh i dinamichekikh charakteristik reaktora 3-go bloka pri energovyrabotke 625 effektivnykh sutok (mart 1982)*, (LAES-protocol No. 15/2833, May 06, 1982)

3.7 V.E. Druzhinin, Yu.A. Tishkin *Tochechnaja metodika rascheta jadernoj bezopasnosti reaktorov RBMK*, (Voprosy atomnoj nauki i techniki, ser. fiz. i tekh. jadernykh reaktorov, Vol.2, p. 3, 1986)

3.8 *Raschetnyj analiz eksperimentov na reaktore 3-go bloka Chernobylskoj AES*, (IAE-Report No. 33R/I-294-88 DSP, 1988)

3.9 *Issledovanije charakteristik reaktorov RBMK s modernizirovannoj grafi-tovoj kladkoj*, (IAE-Report No. 33R/I-524-89, 1989)

3.10 *Fizicheskij pusk reaktora RBMK-1000 3-go bloka Chernobylskoj AES posle dlitelnoj ostanovki*, (ChNPS-Report No. 199-PTO, Dec. 05, 1987)

3.11 *Protokol resultatov izmerenij podkritichnosti i effekta obezvozhivaniya KO-SUZ na reaktore vtorogo bloka Leningradskoj AES v period s 04.06.89 po 09.06.89*, (LNPS-Report, June 9, 1989)

3.12 *Po rezultatam eksperimentov po otsenke izmereniya fizicheskikh kharak-teristik aktivnoj zony reaktora vtorogo bloka ChAES v rezultate zameny II sterzhnej USP na sterzhni RR*, (ChNPS-Report No. 22-1760, April 28, 1989)

3.13 *O vliyanii konstruktsij sterzhnej SUZ na koefficienty i effekty reaktivnosti reaktora RBMK*, (IAE-Report No. 33R/I-525-89, 1989)

3.14 *Protokol rezultatov eksperimentalnogo opredelenija effektivnosti sterzhnej SUZ na reaktorakh 1 i 2 blokov LAES imeni W.I. Lenina*, (LNPS-Report No. 42-03/01-597, February 08, 1989)

3.15 *Tekhnicheskoje reshenije No 12-011 o zamene 12 DKEV, raspolozhennykh v kanalakh SUZ, na 12 DKEV, ustanavlivaemykh v TVS cb. 49 na pervykh ocheredyakh reaktorov RBMK-1000*, (NIIKIET-Report No. TR.120-0137, iskh. No. 120-02/5620, May 31, 1989)

3.16 J.R. Ackew, F.J. Fayers, P.B. Kemshell: *A general description of the lattice code WIN*, J. Brit. Nucl. Energy Soc., **5** pp.564, 1966

3.17 D. Bell, S. Gleston in *Teoriya jadernykh reaktorov*, (Atomizdat, Moscow 1974) p. 496

3.18 V.S. Romanenko, A.V. Krajushkin in *Raschet vybega nejtronnoj moshtnosti v RBMK-1500 pri avarii s poterej teplonositelya*, (Kurchatov IAE-Report No. 33/438583, 1983)

Chapter 6

6.1 N.D. Tarakanov: *Chernobylskie zapiski ili razdymaya o nravstvennocti*, (Voenizdat, Moscow 1989) p. 208

Chapter 9

9.1 Ye.I. Chazov, L.A. Il'yin, A.K. Guskov: *The Dangers of Nuclear War*, (A.P.P., Moscow 1982) p.121

9.2 U.Ya. Margulis: *Atomic Energy and Radiation Safety*, (Energoatomizdat, Moscow 1988) p.91; Table 4.4

Appendix C

C.1 N.S. Babaev, V.F. Demin, L.A. Il'yin, V.A. Knizhnikov, I.I. Kuz'min, V.A. Legasov, Yu.V. Sivintsev: *Yadernaya energetica, chelovek i okruzhayushchaya sreda*, (Energoatomizdat, Moscow 1984)

C.2 L.A. Buldakov et al.: *Radiatsionnaya bezopasnost' v atomnoy energetike*, (Atomizdat, Moscow 1981)

C.3 A.S. Nikiforov et al.: *Obezvrezhivanie zhidkikh radioactivnykh otkhodov*, (Energoatomizdat, Moscow 1985)

C.4 A.N. Marey et al.: *Radiatsionnaya kommunal'naya gigiena*, (Energoatomizdat, Moscow 1984)

C.5 J. Rust, L. Willer (Eds.): *Bezopasnost' yadernoy energetiki*, (Atomizdat, Moscow 1980)

C.6 V.F. Dritschko et al.: *Fragen der Normierung der Konzentrationen von Isotopen der Uran- und Thoriumreihen in Phosphatdüngern* in *Rep. Staatl. Amtes Atomsicherheit und Strahlenschutz der DDR*, **280**, 142–147 (1981)

C.7   *Istochniki i deystvie ioniziruyushchei radiatsii*, Scientific committee of the United Nations, Volumes 1–3, talks presented in 1977 (United Nations, New York 1978)

C.8   U.Ya. Margulis: *Atomnaya energetika i radiatsionnaya bezopasnost'*, (Energoatomizdat, Moscow 1988)

C.9   *Ionizing Radiation: Sources and Biological Effects*, (Unscear 1982); Report to the General Assembly, UN 1982

C.10 *Svodnyi tom materialov po MOYaTTs*, (IAEA, Vienna 1980)

C.11 B.M. Andreev, S.F. Medovshchikov, V.V. Frunze, A.I. Shafiev: *Tritii i okruzhayushchaya sreda*; (Review, Moscow 1984)

C.12 Doklad NKDAR. Vol. 1 *Prilozhenie V. Estestvennye istochniki ioniziruyushchikh izluchenii*, Doklad za 1977. General Assembly of the United Nations (United Nations, New York 1978)

C.13 E.D. Mordberg et al.: *Perokhod izotopov uran-radievogo ryada v zerno nekotorykh sel' skokhozyaistvennikh kul' tur* in *Gigiena i sanitariya* **2** 58-61 (1976)

List of Names

List of Maps, Sketches, and Diagrams

List of Tables

Subject Index

reservoir

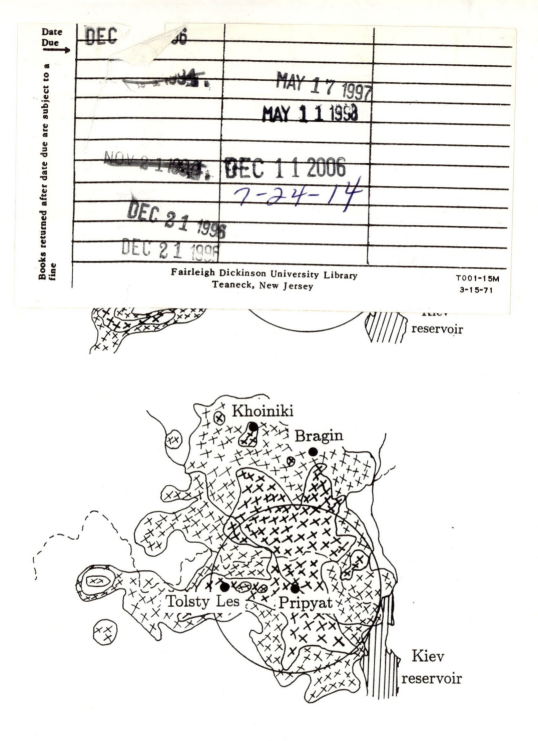

Top: cesium-137
Bottom: strontium-90

Light (dark) areas correspond
to 40 – 110 (more than 110) kBq/m²

Contamination maps adopted from: